BANNED

BANNED

A History of Pesticides
and the Science of Toxicology

Frederick Rowe Davis

Yale
UNIVERSITY PRESS

New Haven & London

Published with assistance from the
foundation established in memory of
Philip Hamilton McMillan of the Class
of 1894, Yale College.

Copyright © 2014 by Frederick Rowe Davis.
All rights reserved.
This book may not be reproduced, in
whole or in part, including illustrations, in
any form (beyond that copying permitted
by Sections 107 and 108 of the U.S.
Copyright Law and except by reviewers
for the public press), without written
permission from the publishers.

Yale University Press books may be
purchased in quantity for educational,
business, or promotional use. For
information, please e-mail
sales.press@yale.edu (U.S. office)
or sales@yaleup.co.uk (U.K. office).

Set in Galliard with Poetica Display type
by Westchester Book Group.
Printed in the United States of America.

ISBN: 978-0-300-20517-6 (cloth)

Library of Congress Control Number:
2014944393

A catalogue record for this book is
available from the British Library.

This paper meets the requirements of
ANSI/NISO Z39.48-1992
(Permanence of Paper).

10 9 8 7 6 5 4 3 2 1

To my parents
Dan Clifford Davis
and
Judith Rowe Davis

. . .

When the public protests, confronted with some obvious evidence of damaging results of pesticide applications, it is fed little tranquilizing pills of half truth. We urgently need an end to these false assurances, to the sugar coating of unpalatable facts. *It is the public that is being asked to assume the risks that the insect controllers calculate.* The public must decide whether it wishes to continue on the present road, and it can do so only when in full possession of the facts.

—Rachel Carson, *Silent Spring*

. . .

CONTENTS

Preface, ix

Acknowledgments, xv

List of Abbreviations, xix

CHAPTER 1.
Toxicology Emerges in Public Health Crises, 1

CHAPTER 2.
DDT and Environmental Toxicology, 38

CHAPTER 3.
The University of Chicago Toxicity Laboratory, 72

CHAPTER 4.
The Toxicity of Organophosphate Chemicals, 91

CHAPTER 5.
What's the Risk? Legislators and Scientists
Evaluate Pesticides, 116

CHAPTER 6.
Rereading *Silent Spring*, 153

CONTENTS

CHAPTER 7.
Pesticides and Toxicology after the DDT Ban, 187

CHAPTER 8.
Roads Taken, 214

Epilogue: Risk, Benefit, and Uncertainty, 221

Notes, 225

Index, 253

PREFACE

Late in 1972, William Ruckelshaus, administrator of the U.S. Environmental Protection Agency, announced that he was canceling the registration for DDT, in effect banning the use in the United States of one of the most popular insecticides available since its introduction after World War II. Environmentalists hailed the ban on DDT as a crowning achievement of the American environmental movement and the culmination of a decade of activism that had coalesced in the wake of the publication of *Silent Spring* by Rachel Carson in 1962. Carson's damning critique of the indiscriminate use of chemical pesticides and the resulting widespread ecological contamination in the United States captured the hearts and minds of Americans as few other books had, and it inspired extensive hearings in the President's Science Advisory Committee and Congress. The passage of the National Environmental Protection Act in 1970 and the establishment of the Environmental Protection Agency (EPA) that same year signaled to Americans that their concerns had been heard. The DDT ban terminated applications of one of the most notorious and environmentally destructive chemicals in America. Could there be a more perfect conclusion to a dark chapter in the story of American agriculture and public health?

In May 1982, several bird-watching friends (retirees) invited me to join them near Rochester, New York, to find as many bird species as we could in a single day. The big day started at about 1:00 a.m. as we headed out to find nocturnal owls and nighthawks. By 4:30, we had arrived at Norway Road, a renowned hotspot for migrating birds west of Rochester. In the early morning darkness, we heard an American woodcock,

ix

PREFACE

and just before dawn a veery began singing its ethereal song and a wood thrush soon joined in. By daybreak, dozens of other species—neotropical migrants—could be heard: warblers, vireos, thrushes, tanagers, cuckoos, flycatchers, and sparrows. The phrase "dawn chorus" fails to capture the deafening cacophony of thousands of birds in full breeding song. My ears were still ringing when I arrived back home near midnight. Even as a teenaged birder I knew that I had witnessed something special that morning. It did not occur to me that I would never again hear anything like the dawn chorus I heard on the May morning so long ago.

Fast-forward thirty years to March 12, 2013, the *New York Times* reported on a study revealing that acutely toxic pesticides correlated to declines in grassland birds in the United States more closely than habitat loss through agricultural intensification. Three days later, a federal judge considered a motion to dismiss a 2011 lawsuit in which the Center for Biological Diversity and the Pesticide Action Network alleged that the EPA has allowed pesticide use without required consultations with federal agencies. What is going on here? More than fifty years after the publication of *Silent Spring* and forty years after the ban on DDT, pesticides still account for widespread deaths in birds and other wildlife. Could it be that Carson's statement in the epigraph applies to the risks of pesticides today? My purpose in writing *Banned* is to explain one the great ironies of the American environmental movement.

Writing this book emerged out of a simple objective. I had read Rachel Carson's *Silent Spring,* and I wanted to explore her scientific and medical sources in building the case against DDT. As I originally understood the story, *Silent Spring* was a significant catalyst in the environmental movement of the 1960s and 1970s as well as in the banning of DDT. But I soon discovered that Carson began researching pesticides in earnest during the late 1950s, at a time when American scientists, legislators, and regulators were debating the risks and benefits of the still-new chemicals as they revised existing laws. Moreover, concerns about chemical exposures had much deeper roots that significantly predated this period.

Inspired by the fiftieth anniversary of the publication of Carson's book, *Banned* examines the development of synthetic pesticides and the science of toxicology over the course of the twentieth century, as well as

x

PREFACE

the legislation that governs exposures to these chemicals—from the Pure Food and Drug Act of 1906 to the Food Quality Protection Act of 1996 and beyond. By studying the larger context of the evolution of pesticides and toxicology, in this book I reveal how environmental science and policy evolved across the twentieth century. Carson published *Silent Spring* at a critical moment in the history of environmental science. Studying her sources and rereading her careful interpretations of science deepens our appreciation of the achievement of *Silent Spring*. But an analysis of pesticides policy and usage patterns in the decades following the ban on DDT exposes a tragic irony in the worldwide proliferation of the organophosphate pesticides, chemicals that Carson and the toxicologists acknowledged to be far more toxic to wildlife and humans alike.

In *Silent Spring*, Carson cited numerous cases of acute and chronic pesticide poisonings in wildlife, ecosystems, and humans. Throughout her book, she implored people to evaluate evidence and assess threats to environmental and public health, including their own exposures. Carson exhaustively reviewed the scientific and medical literature on how synthetic insecticides affect humans and the environment. To clarify her indictment of the chemical industry and federal agricultural and public health programs, she dramatized scientific and medical findings and thus gave them a face. As much as *Silent Spring* was about toxicity and lethal doses, it was also about the victims of poisonings: farm workers, children, American robins, bald eagles, and Atlantic salmon. Moreover, Carson revealed how pesticides like DDT permeated soil and waterways all across America.

In part, Carson blamed overspecialization for the lack of knowledge regarding ecological and health effects: each specialist focused on his or her own problem, oblivious to its larger frame. Carson introduced Americans to an emergent discipline and brought the language of environmental toxicology into the public realm. Phrases and concepts like "acute and chronic toxicity," "LD_{50}," "parts per million," "carcinogenicity," and "reproductive effects" came to dominate the study and regulation of environmental risks of synthetic insecticides. And in so doing, *Silent Spring* alerted the American public to the problems of indiscriminate use of pesticides and the science that sought to assess the risks they imposed by using the language of toxicology.

xi

PREFACE

Carson's indictment of pesticides instigated heated debate. Representatives of the chemical industry challenged her authority and the scientific legitimacy of the book. Some of the attacks became personal and attempted to dismiss Carson on the basis of her gender. Even among scientists without direct ties to the chemical industry, reviews of *Silent Spring* were mixed. Certain scientists praised Carson for the breadth and depth of her analysis, but others took issue with details and Carson's broad conclusions. President John F. Kennedy directed the President's Science Advisory Committee (PSAC) to convene hearings and report on the risks and benefits of pesticides, and congressional hearings soon followed. Nevertheless, the wheels of bureaucracy ground slowly. With the exception of minor adjustments to existing laws, Congress failed to pass significant legislation. Finally, litigation at the state and federal levels prompted the EPA to ban use of DDT in the U.S. late in 1972, though manufacture and exportation of DDT continued.

Just as exploring the sources of *Silent Spring* revealed its roots in science and policy, analysis of developments in its wake uncovered a tragic irony. The story of the book that launched the environmental movement became much more complicated after the ban on DDT when some of the most toxic chemicals known came to dominate the pesticides market. Although there have been many studies of DDT and the chlorinated hydrocarbons, few scholars have explored the history of organophosphates and the toxicology of this large class of pesticides. For the most part, historians have divided the history of pesticides into three periods: pesticides before DDT, pesticides during the DDT era, and pesticides after DDT (still largely unexamined).

In contrast, in *Banned* I evaluate chemical insecticides by chemical class, like a chemist, and by toxic insult, like a toxicologist. It may surprise readers to learn that most of the pesticides that predominated in agriculture and public health during the twentieth century can be divided into a few major classes: heavy metal insecticides, such as lead arsenate and Paris green; chlorinated hydrocarbons like DDT; organophosphates (and carbamates), such as parathion and carbaryl; synthetic pyrethroids like permethrin; and even the now-prolific neonicotinoids. Environmental scientist John Wargo argued that the chemical-by-chemical approach delayed both toxicological analysis and effective legislation, not to mention historical analysis. By focusing on classes of

xii

PREFACE

chemicals across the twentieth century (which is to say before, during, and after the DDT era), I sought underlying patterns to understand the development of toxicology and legislation.

As I studied the history of pesticides, the publication of *Silent Spring* emerged as a critical moment in a century of attempts by scientists, legislators, and corporations to manage their risks and benefits. The complexities of the topic show us how rigorous and thoughtful Carson was in her writings about pesticides. By relocating *Silent Spring* within the broader context of the science and regulation of pesticides, I reveal both the inspiration and the implications of one of the most important studies of our time.

Banned reveals the evolution of the science of toxicology and the development of pesticides in the United States during the twentieth century. Since publication of *Silent Spring*, toxicology and environmental risk assessment have become the dominant paradigms for how scientists determine environmental damage and threats to the health of humans, wildlife, and ecosystems. Environmental toxicology developed out of the science of pharmacology. Toxicologists, many of whom initially studied pharmacology, accepted the fundamental aphorism of toxicology, attributed to Paracelsus: "The right dose differentiates a poison and a remedy." The dynamic between risk and benefit inherent in pharmacology permeated the science of toxicology also. But the science of toxicology and its development captures only part of the story of pesticides in modern America. Long before toxicology emerged as an independent discipline, pesticides had become integral to agriculture in America. As agriculture industrialized to a previously unimagined scale, farmers increasingly depended on chemical pesticides to control insects and prevent extensive damage to crops.

Much of the story that follows examines how scientists clarified risks associated with pesticides, but the benefits of chemical insecticides in American agriculture and public health also warrant our attention. Chemical companies were quick to emphasize benefits while minimizing potential risks. The public (both as producers and consumers), however, judged the value of pesticides and other chemicals for themselves. Popular science writers like Rachel Carson interpreted (and interpolated) scientific research and legislative debate for the public. Thus what began as a study of a group of scientists developed into a broader examination

xiii

PREFACE

of the intersections between scientists, policy makers and enforcers, corporate representatives, science writers, and the public.

Banned interweaves several separate narratives. One narrative strand reveals how toxicology evolved as a scientific study. Another follows individuals and groups of scientists and affiliated institutions responsible for the toxicological evaluation of pesticides and other chemicals. To the extent possible, I reflect on the culture of toxicology as fostered by such leaders as E. M. K. Geiling at the University of Chicago Toxicity Laboratory (Tox Lab) and Arnold Lehman of the Pharmacology Division at the FDA. The toxicological analysis of an ever-expanding number of novel chemicals provides yet another narrative. Still another examines the development of legislation to regulate pesticides and other potentially toxic substances. All of these separate stories intersected with the publication of *Silent Spring,* which encouraged the public to judge for itself the risks posed by synthetic chemicals in the environment. Ironically, after DDT and other chlorinated hydrocarbons were banned in 1972, farmers turned to the highly toxic organophosphate insecticides as replacements, and the regulatory analysis of this class of chemicals stretched into the new millennium. The spectacular proliferation of the neonicitinoid insecticides in agriculture over the past decade has reignited debate about environmental hazards of pesticides.

Over the course of the twentieth century, scientists, physicians, and public health officials have attempted to characterize the risks posed by insecticides and other chemicals. Yet insects continue to pose a real threat to health and welfare. Farmers, public health officers, and consumers have demanded effective controls for unwanted insects. In the case of pesticides, the tightly bound helix of risk and benefit defies simple solution.

xiv

ACKNOWLEDGMENTS

It is fitting that this book, which had its origins at Yale, has found its way back to that distinguished university. I am very grateful to a large and diverse group of scholars there including the late Frederic L. Holmes, Dan Kevles, John Wargo, John Mack Faragher, Cynthia Russett, Steven Stoll, Toby Appel, Susan Lederer, and John Warner. In addition, I deeply appreciate the assistance of many librarians at the Beinecke Rare Book and Manuscripts Library, Yale School of Medicine, Sterling Memorial Library, and the Yale School of Forestry and Environmental Studies. A fellowship from the Beinecke Library facilitated a summer of research in the Rachel Carson Papers. The following statements serve as grateful acknowledgment of permissions to publish quotations from *Silent Spring*. Quotations from *Silent Spring* by Rachel Carson (Copyright © 1962 by Rachel L. Carson) are reprinted by permission of Frances Collin, Trustee, and Houghton Mifflin Harcourt. Quotations from *Silent Spring* by Rachel Carson (Copyright © 1962 by Rachel L. Carson, renewed 1990 by Roger Christie) are reprinted by permission of Houghton Mifflin Harcourt Publishing Company. All rights reserved. I also acknowledge Magnum Photos for permission to reprint Erich Hartmann's photograph of Rachel Carson. Additional photographs of E. M. K. Geiling, Kenneth DuBois, and John Doull appear courtesy of the Special Collections Research Center, University of Chicago Library. Special thanks go to Christine Coburn for providing access to these photos and to Daniel Meyer for authorizing permission to publish them.

My current university, Florida State University, has been a very conducive place to complete this study. I am particularly grateful to

xv

ACKNOWLEDGMENTS

colleagues both current and former, including Neil Jumonville, Edward
Gray, Jonathan Grant, Elna Green, Michael Ruse, Michael Creswell,
Richard Mizelle, Joe Gabriel, Andrew Frank, Jennifer Koslow, Peter
Garretson, Charles Upchurch, Ron Doel, Kristine Harper, Max Fried-
man, and Will Hanley. The History Department provided funds for travel
to research sites and conferences. The Office of Sponsored Research at
FSU gave me three generous awards for travel, research, and writing,
including the Developing Scholar's Award. A Grant for Scholarly Works
in Biomedicine (#1G13LM009606) from the National Library of Medi-
cine at the National Institutes of Health provided a much-needed release from teaching responsibilities to complete research and writing.
With this grant, I was also able to utilize a full-year sabbatical from
FSU. A small army of graduate students contributed in numerous ways
to the research and editing of this book as readers and research assistants:
Michael Bonura, Gary DeSantis, Elizabeth Dobson, Abraham Gibson,
Samantha Muka, Darryl Myers, Samiparna Samanta, Harvey Whitney, and
Chris Wilhelm.

Audiences at a wide range of conferences and workshops have re-
sponded to presentations drawn from this research. These include the
American Society for Environmental History, the History of Science
Society, the Rachel Carson Center in Munich, Germany, Cermes3—
INSERM/CNRS, RiTME—INRA with support from the French National Research Program for Endocrine Disruptors, the Agricultural
History Society, Department of Biology at Washington University,
Lyman Briggs College at Michigan State University, and the College
of Life Sciences at University of Missouri. Conversations with authorities too numerous to list have shaped my thinking about pesticides and
toxicology as well as *Silent Spring*. This is a partial list: Roland Clement, John Doull, Linda Lear, Mark Lytle, Michelle Mart, Gary Mor-
mino, and Richard Pough.

Numerous erudite scholars read versions of the manuscript or sections thereof. I am very grateful for the detailed comments of Mark Bar-
row, Mark Borello, David Hecht, Frederic Holmes, Dan Kevles, Nancy
Langston, Michael Ruse, John Wargo, and Judith and Dan Davis, as
well as anonymous reviewers. I hasten to note that any remaining errors
of fact or judgment are mine alone. At Yale University Press, I thank
Jean Thomson Black, Samantha Ostrowski, and Eliza Childs.

xvi

ACKNOWLEDGMENTS

Family and friends have supported me in countless way. Again, I am grateful to my sister and brothers and their families. The imminent birth of my son Spenser Lake Davis provided the impetus to finish the first draft, and in what seems like the blink of an eye, I can discuss the book with him in detail. Spenser also devised the most alliterative suggested title. Words cannot express the gratitude I feel for my parents. Better than any other source, my father's boyhood recollections captured the wonder of DDT to millions of American farmers, and my mother shared her love of environmental literature. My parents have always supported and shared my passion for birds and the environment. I cherish our collective memories of many spring mornings spent in search of warblers and other migratory birds. Here's to next spring, Mom, Dad, and Spenser!

ABBREVIATIONS

AEC	Atomic Energy Commission (AEC)
AMA	American Medical Association
AOAC	Association of Official Agricultural Chemists
BEPQ	Bureau of Entomology and Plant Quarantine
BHC	benzene hexachloride
CMR	Committee on Medical Research
DDT	dichloro-diphenyl-trichloroethane
DFP	diisopropyl fluorophosphate
EPA	Environmental Protection Agency
FDA	Food and Drug Administration
FEPCA	Federal Environmental Pesticide Control Act
FFDCA	Federal Food, Drug, and Cosmetic Act
FIFRA	Federal Insecticide, Fungicide, and Rodenticide Act
FWS	United States Fish and Wildlife Service
HETP	hexaethyl tetraphosphate
HEW	Department of Health, Education, and Welfare
LD_{50}	Lethal Dose 50 / Median Lethal Dose
mg/kg	milligrams/kilogram
µg/kg	micrograms/kilogram
NDRC	National Defense Research Committee

xix

ABBREVIATIONS

NIH	National Institutes of Health
OMPA or Pestox III	octamethyl pyrophosphoramide
OSRD	Office of Scientific Research and Development
PFDA	Pure Food and Drug Act
PHS	United States Public Health Service
ppb	parts per billion
ppm	parts per million
PSAC	President's Scientific Advisory Committee
TOCP	triorthocresyl phosphate
Tox Lab	University of Chicago Toxicity Laboratory
USDA	United States Department of Agriculture

CHAPTER 1

Toxicology Emerges in
Public Health Crises

In 1893 the city of Chicago announced to the world its arrival as a major metropolis through the Columbian Exposition. That same year, the British medical journal *Lancet* deployed a public health doctor to report on the state of the Chicago stockyards, which in terms of size and productivity were among the largest in the world. In the Chicago stockyards, slaughterhouses became industrial meat factories that rendered thousands of cattle into meat products to be transported via refrigerated railcar across the United States and shipped around the world. With the industrialization of meat production came spectacular demand, and the Chicago packinghouses refined meat production to the essence of efficiency. Nothing mattered save enhanced productivity and profit. Safeguards fell by the wayside, endangering workers and consumers alike.[1] Technical reports failed to excite concern, but when Upton Sinclair published *The Jungle* in February of 1906, the conditions in the Chicago stockyards caught the attention of legislators, and even the president responded. Having languished in the form of several related bills for nearly two decades, the Pure Food and Drug Act (PFDA) was finally passed in June of 1906. Primarily directed at product labeling, the PFDA prohibited interstate transport of unlawful food and drugs.

The industrial principles of efficiency that infused the Chicago packinghouses extended to other dimensions of agriculture as well. Novel technologies facilitated the expansion of monoculture as the preferred and

1

TOXICOLOGY EMERGES IN PUBLIC HEALTH CRISES

most profitable method, but for the threat of insect invasion. Paris green, and later lead arsenate, provided a technological fix to this problem. Adelynne Whitaker, historian of pesticide legislation, argued that agricultural chemists and entomologists during the early twentieth century were primarily concerned with "the economic aspects of adulterated and ineffective insecticides," but "the scientists were not unaware of the public health implications of their work."[2] Just as the Chicago packers found an insatiable demand for their low-cost meat products, farmers found virtually unlimited demand for their produce. In both cases, industrializing modes of production resulted in previously unimagined levels of productivity. As production became concentrated in Chicago and points farther west, Americans increasingly found themselves separated from the sources of their food. The PFDA and the Insecticide Act of 1910 reassured Americans that despite such detachment they could still expect a safe and healthy food supply. During the next three decades the foundations of such presumptions were shaken as Americans found their health and welfare threatened by pesticides and adulterated drugs. Yet it took a national tragedy to shake American consumers and regulators out of their complacency.

There is no question that the PFDA represented a watershed moment in the history of regulation in the United States. Yet legislators recognized that the act incorporated compromises that left American public health vulnerable to corporate deceit and malfeasance. By the 1930s, the president had called on Congress to revisit food and drug legislation and forge a law with greater power to protect Americans. As the House and the Senate debated health legislation, several crises accentuated the limitations of the 1906 law, notably its inability to prosecute abuses that led to injury and even death. In addition to legislators, a new generation of consumer advocates took up the cause, writing books and articles to alert Americans to the failings of existing legislation and to the flagrant violations that exposed them to significant risks. Among the many examples that failed to elicit regulation, the case of Jamaica ginger ("ginger jake") paralysis, in which many thousands were poisoned after unwittingly consuming a highly toxic alternative to alcohol during Prohibition, was particularly egregious. The pesticide residues of arsenic and lead on fruit also failed to motivate consumers or legislators. Both ginger jake and pesticide residues received considerable coverage in the

2

media (and in the books and articles of consumer advocates), but not until the Elixir Sulfanilamide tragedy, in which ninety-three individuals died after ingesting a contaminated drug, did legislators pass the Federal Food, Drug, and Cosmetic Act of 1938 (FFDCA), thereby revising the 1906 law.

The Elixir Sulfanilamide tragedy, similar incidents, and the legislation that followed gave toxicology new standards and food and drug laws increased leverage.[3] Moreover, several scientists and regulators launched their lengthy careers with the study of Elixir Sulfanilamide. Through their work on Elixir Sulfanilamide they learned vital lessons and developed new approaches to toxicology that they would continue to apply to pharmaceuticals and other new chemicals, such as insecticides, throughout their careers in government, industry, and research universities. Such cases also affected popular perceptions of risk in America. In addition to an expectation that government would provide safety standards for pharmaceuticals, Americans were becoming accustomed to the use of powerful chemicals in medicine as therapies. Similarly, new technologies, including chemical insecticides, accelerated the industrialization of agriculture in America during the first decades of the twentieth century.

The scale of meatpacking drove intensification of production, but meat was not the only agricultural commodity that was industrialized during the last half of the nineteenth century. The agricultural revolution saw the introduction of new technology in form of steel plows, seed drills, cultivators, and reapers, which greatly reduced the need for a large labor force. Moreover, methods of crop rotation and the application of fertilizers significantly enlarged yields of many crops. Mechanization and fertilization meant that established farmers could plant extensive crops consisting of monocultures. Successful harvests could be spectacularly profitable. Nevertheless, monoculture left crops profoundly vulnerable to insect invasions, which could quickly bankrupt ambitious farmers. Historian James Whorton placed these agricultural developments in context: "The favorable insect environment created by monoculture was further enhanced in America by westward expansion. The fulfillment of Manifest Destiny not only involved an enormous increase in the area of land under cultivation but, also, by the prerequisite clearing of forests in many areas, frequently destroyed predators of insects while

TOXICOLOGY EMERGES IN PUBLIC HEALTH CRISES

forcing the insects themselves to turn to a domestic food supply."[4] Even
the novel technologies, such as railroads and trans-Atlantic ships, that facilitated a related revolution in transportation contributed to the problem
of insect invasions by transferring the culprits around the country.

Farmers became desperate for effective means to control insect infestations. Economic entomologists, hoping to escape unfavorable stereo-
types as ineffectual, disengaged scientists preoccupied with some of the
smallest and most inconsequential members of the animal kingdom,
answered farmers' hopes. Particularly promising was an insecticide
extracted from the pyrethrum flower, a chrysanthemum. Drying and
crushing the stamens of the flowers produced a powerful insecticide.
Pyrethrum, as the insecticide became known, was prohibitively expensive because farmers in the Caucasus guarded their monopoly on the
plant.[5]

Economic entomologists sought a synthetic insecticide as effective as
pyrethrum. In 1867, farmers received the answer to their prayers in the
form of Paris green, a copper acetoarsenite. Some journalists warned
against adding arsenic to agriculture, but farmers soon adopted Paris
green to fight a range of insect pests. In the five years after its introduction, Paris green became "the ally of first resort whenever death must be
dealt to any pest."[6] Paris green was popular with farmers because it was
inexpensive and effective against a variety of insects. The wonder insecticide met its match after a Harvard astronomer with an interest in silk
production imported the gypsy moth, which took flight in 1869. Their
caterpillars soon stripped trees around Medford, Massachusetts, of their
leaves, and the moths expanded their range around New England. Heroic efforts at control included setting caterpillars aflame with kerosene. By 1890, gypsy moth caterpillars threatened orchards and forests
throughout New England.[7] Surprisingly, Paris green failed to control
the resistant caterpillars. In 1892, F. C. Moulton, a Gypsy Moth Commission chemist, had introduced the solution in the form of lead arsenate. Its effectiveness outweighed its expense, and as an added benefit it
was gentler on the foliage on which it was sprayed. At the turn of the
century, lead arsenate had become the preferred insecticide, a position it
would continue to occupy until the introduction of DDT after World
War II. Despite the proliferation of Paris green and lead arsenate, regu-

4

lation of pesticides languished until the passage of PFDA in 1906 and, more important, the Insecticide Act of 1910.

Once the PFDA established labeling standards for drugs and foods in the U.S. in 1906, prospects for comparable standards for insecticides (and fungicides) began to look very promising. E. Dwight Sanderson reported that one of the nation's largest insecticide manufacturers agreed to include an analysis of its goods on all labels. As the director and entomologist of the New Hampshire Experiment Station and the head of the Standing Committee on Proprietary Insecticides of the Association of Economic Entomologists, Sanderson interpreted the agreement as a promising sign that the insecticides industry would support his committee's resolution for national labeling legislation for insecticides and fungicides.[8] Harvey Wiley, director of the Bureau of Chemistry and a staunch advocate for the regulation of foods and drugs, encouraged Sanderson to lobby Congress for such legislation for three reasons: the PFDA did not extend to insecticides, an amendment to the PFDA was not feasible, and labeling of insecticides was critically important. When compared to the tortuous path of the food and drug act, passage of the insecticide act was both speedier and more direct. The passage of the Insecticide Act in April 1910 confirmed Sanderson's impression that the insecticide industry was ready for regulation. Historian Adelynne Whitaker argued that the quick passage of Insecticide Act hinged on industry acceptance, but insecticides producers witnessed how the food and drug act benefited responsible and reliable producers of food and drugs, whereas producers of adulterated goods were forced out of the market.[9]

As the PFDA set labeling standards for foods and drugs, the Insecticide Act of 1910 established similar standards for insecticides. The manufacture and sale of adulterated and misbranded insecticides and fungicides in interstate commerce became illegal under the insecticide law. It codified legislative standards for insecticides in general and specifically the two insecticides that were most commonly used in the United States: lead arsenate and Paris green. Regulators expected specific standards to reduce problems with enforcement, which had posed difficulties for the USDA.[10] Enforcement of the federal Insecticide Act fell to the Insecticide and Fungicide Board within the Bureau of Chemistry. Early in the act's history, few if any noted the potential conflict of

TOXICOLOGY EMERGES IN PUBLIC HEALTH CRISES

interest, but by the mid-1920s, the USDA's dual role as protector of both farmers and consumers garnered criticism, particularly as the problem of pesticide residues on produce became a matter for regulatory concern.[11] Understanding of the risks posed by heavy metals contaminants, such as lead and arsenic, owed much to the pioneering work of a few researchers who were establishing the new study of industrial hygiene.

The history of industrial hygiene (later, occupational medicine) owes much to the sweeping research of the social historian of medicine Christopher Sellers.[12] Other historians have made important contributions with respect to specific toxins, diseases, and companies.[13] In general, the U.S. lagged behind Britain and Germany in the evolution of industrial hygiene. Sellers attributed the lag to several factors, including the ambiguity of indeterminate symptomology of industrial diseases, delayed onset of diseases, a worker culture that dictated stoicism and personal fortitude in the face of workplace hazards, rapid turnover in employment, fear and mistrust of company physicians and orthodox medicine in general, financial barriers to medical care, and a general lack of knowledge of industrial diseases (exacerbated by the increasing occupational specialization).[14] In Great Britain and Germany, pioneer researchers in academic or government posts began to study occupational disease in the late nineteenth century. Particularly noteworthy were the efforts of Karl Lehmann as the director of the Hygienic Institute in Wurzburg. Lehmann began studies the effects of gases and vapors on cats.[15]

In the U.S., systematic study of occupational disease took form in Alice Hamilton's studies of working conditions in Illinois factories beginning in 1910. Hamilton (1869–1970) meticulously documented the hazards of various occupations, most notably the lead industry. Remarkably, without specific authority, she depended on the good faith of lead companies to allow her to survey workers and determine their illnesses and the causes thereof. After surveying several companies, Hamilton was able to compare lead poisoning rates among their workers. Such comparison led to informal competition between companies as they tried to lower their disease rates below those of their rivals. Hamilton later expressed mixed feelings regarding the effect of such competition and hoped that the companies sought the moral high ground rather than favorable cost-benefit ratios.[16]

TOXICOLOGY EMERGES IN PUBLIC HEALTH CRISES

Hamilton's research documented effects of lead poisoning, but it remained for other researchers to determine pathways of toxic insult. In 1921, several researchers at Harvard Medical School, where Hamilton had joined the faculty as professor, launched what became known as the "lead study." Cecil Drinker and Joseph Aub, the Harvard faculty members who led the lead study, envisioned a study of occupation disease grounded in science. Both strove to distinguish their work from Hamilton's socially conscious efforts. Although David Edsall, Drinker, and Hamilton negotiated with the lead companies for "full scientific liberty," Aub never forgot his obligations to the companies he studied, even as he developed a clinical and scientific research agenda. In a critical early move, Aub called upon a chemist—Lawrence Fairhall—to develop a more reliable method of analyzing minute amounts of lead in biological material. In addition to the chemist, Aub would soon incorporate a pathologist and physiologists into the research of the lead study, which anticipated the spirit of interdisciplinary collaboration that exemplified the study of toxicology. Moreover, Sellers argued, the lead study placed the toxicological approach at the core of studies of industrial disease.[17] The Harvard study shaped the toxicological approach to the study of industrial disease in other important ways. Fairhall's method of quantitative analysis provided a uniform basis for comparing lead levels in workers to the levels in their surroundings. Researchers could compare lead levels within factories and between industries. With the development of accurate techniques of analyzing other chemical hazards, dangers could be placed along single quantitative scales, typically in the same units (milligrams per ten cubic meters). Aub and the Harvard researchers also incorporated laboratory experiments into the lead study. These experiments, using both humans (volunteers) and animals, enabled the researchers to isolated specific workplace causes and effects. Such experiments transferred the research out of the workplace, to laboratories where researchers could study the more dangerous effects of lead and other toxins on cats or rabbits including pathological examination, which was unthinkable for humans, even volunteers. Finally, even as Aub and the other researchers in the Harvard lead study shifted the locus of the study from the workplace to the laboratory, they maintained strong ties to managers and doctors at some of the largest corporations in the U.S. Sellers

7

TOXICOLOGY EMERGES IN PUBLIC HEALTH CRISES

argued somewhat ironically that the independent course of their research depended on continued corporate support.[18]

One of the first corporations to develop its own toxicological laboratory happened to be the largest chemical company in the U.S.: DuPont. Wilhelm C. Hueper (1894–1978) lobbied for what became the Haskell Laboratory at DuPont. Hueper made many important contributions to the study of occupational and environmental carcinogenicity over the course of a long career.[19] After completing his medical education in Germany, Hueper immigrated to the United States, where he held various posts in academia, industry, and government. As chief pathologist at the University of Pennsylvania's Cancer Research Laboratory and director of pathology at the American Oncologic Hospital, he asked to visit the DuPont Dye Works at Deepwater, New Jersey. At the factory he discovered in use aromatic amines that he knew, based on his experience in German factories, to cause bladder cancer. Through channels, Hueper notified Irénée du Pont, vice chairman of the board, of this potential hazard and recommended that DuPont establish an in-house biological laboratory to conduct toxicity studies. Internally, George Gehrmann advocated for a new lab, and DuPont's Haskell Laboratory open on January 22, 1935, under the directorship of Wolfgang F. von Oettingen. Hueper joined Haskell after the University of Pennsylvania declined to renew his contract in the spring of 1934. Dow and Union Carbide established similar laboratories in the same year.[20]

Oettingen set as the first priority to determine the mechanism of the formation of bladder tumors. By late 1935, seventy DuPont workers had developed bladder tumors, and Hueper and his colleagues Frank Wiley and Humphrey D. Wolfe launched a long-term experimental study of dogs that were fed chemicals, including beta-naphthylamine ("beta"). After three years, the beta-fed dogs developed tumors, whereas those fed the other chemicals remained tumor free. Hueper and his colleagues published their findings, much to the chagrin of the DuPont board, which soon barred Hueper and other Haskell scientists from publishing their results. In a productive three years, Hueper conducted animal studies on a range of DuPont products, including seed grain vermicides, carbon disulfide, ethylene glycol and related solvents, refrigerant gases such as Freon, and Teflon coatings for kitchen utensils, but DuPont management fired him in November 1937 and stipulated that he never

8

publish his research from Haskell. Hueper refused to comply with Du-Pont's stipulation. Historian Robert Proctor has argued that, as a result of his refusal, DuPont hounded Hueper for the rest of his career, even threatening him with a lawsuit when he was invited to speak before the International Union against Cancer.[21]

Nevertheless, Hueper set to work on his magnum opus in 1938 and published his authoritative *Occupational Tumors and Allied Diseases* in 1942. Although Hueper recognized that acute poisonings represented a significant problem in the workplace, he argued that long-term chronic effects, like cancer, were far more important. But since chronic effects often did not manifest until months or years after exposure, physicians might not consider the particular cause of a particular disorder. In addition, Hueper suggested that most occupational cancers were preventable with proper procedures to protect workers, but he doubted that society would accept the necessary precautions to prevent cancer in the workplace.[22] Still, support and participation of U.S. corporations proved to be critical to the development of toxicology and, as Sellers has shown, the toxicological approach to occupational medicine. Notwithstanding these developments in corporations, regulators and consumer advocates found recent legislation lacking.

Despite the general sense of satisfaction with the passage of the 1906 food and drug law and the 1910 insecticide law, as well as the advances in toxicology emerging from research in industrial hygiene, regulators soon confronted the limitations of the legislation. By 1933, several distinct groups questioned the efficacy of the laws. As chief of the Food and Drug Administration, Walter Campbell struggled with the law, particularly because it often fell to him to explain its deficiencies. In one case, the assistant secretary of agriculture, Rexford Tugwell, returned a routine spray-residue letter with a question along the lines of, if lead arsenate was a poison, why didn't the FDA prohibit its use? Reflecting on the moment years later, Campbell's assistant, Paul Dunbar, wrote: "The effect on all of us after these long years of fighting a lone battle against spray residues was like a kick in the teeth."[23] No one was more certain of the FDA's inability to ban lead arsenate based on the 1906 law than its chief. Regulators could not address adequately cosmetics, patent medicines, adulteration of food, and even false advertising under the provisions of the current statute.[24]

TOXICOLOGY EMERGES IN PUBLIC HEALTH CRISES

In the 1930s a new generation of consumer advocates emerged, voicing a sharp critique of the inadequacy of federal food and drug law. In 1933, Arthur Kallet and F. J. Schlink, both with Consumers' Research, Inc., wrote *100,000,000 Guinea Pigs: Dangers in Everyday Foods, Drugs, and Cosmetics,* which constituted a broad indictment of the gaps in existing policy. The title suggested that food, drug, and cosmetic producers treated the 100,000,000 Americans like guinea pigs by exposing them to unknown and unrecognized risks. Kallet and Schlink cited numerous cases that the FDA could not (or would not) prosecute, ranging from food residues to prescription drugs to cosmetics, not to mention cases of false advertising. They split their critique between the 1906 law and the companies that flouted the spirit (if not the letter) of the law. They wrote: "Using the feeble and ineffective pure food and drug laws as a smokescreen, the food and drug industries have been systematically bombarding us with falsehoods about purity, healthfulness, and safety of their products, while they have been making profits by experimenting on us with poisons, irritants, harmful chemical preservatives, and dangerous drugs."[25]

The book *100,000,000 Guinea Pigs* became a model for other consumer advocates, among them Ruth deForest Lamb, who published *American Chamber of Horrors: The Truth about Food and Drugs* in 1936. Lamb ratcheted up the level of concern, focusing particularly on American housewives and their families. In her first chapter, "Why Doesn't the Government Do Something about It?," Lamb commanded her readers' attention right from the opening paragraph: "You've been told you take your life in your mouth every time you bite into an apple or brush your teeth. All of your food is injurious, and your drugs and cosmetics are dripping with poisons. Anesthetic ether is always adulterated, and the ergot on which physicians depend to stop the hemorrhages of childbirth is impotent—unless, of course, it comes from Spain."[26] Thus she asserted that American consumers faced grave dangers through callous abuses on the part of the companies that produced foods, drugs, and cosmetics. Residues of pesticides on fruit were among the most worrisome to scientists and advocates.

As we have seen, chemical insecticides such as lead arsenate and Paris green proliferated during the nineteenth century. Use of these heavy metal insecticides rose dramatically during the first three decades of the twentieth century (table 1). Agricultural applications of arsenates qua-

10

TOXICOLOGY EMERGES IN PUBLIC HEALTH CRISES

Table 1

Insecticide Use (estimated) in the U.S., in Pounds,
1919, 1923, and 1929

	1919	1923	1929
Lead arsenate	11,500,000	11,000,000	29,000,000
Calcium arsenate	3,000,000	31,000,000	29,000,000
Paris green		3,000,000	
Total insecticide use	14,500,000	45,000,000	58,000,000

Source: C. N. Myers, Binford Throne, Florence Gustafson, and Jerome Kingsbury, "Significance and Danger of Spray Residue," *Industrial and Engineering Chemistry* (June 1933): 624.

drupled in the decade between 1919 and 1929. The arsenates appealed to farmers as broad-spectrum insecticides, which is to say, they were very effective against a wide range of insects. In *Before Silent Spring,* Whorton analyzed the toxicological studies of insecticide residues conducted between 1900 and 1920. Entomologists found themselves caught in a tug of war between the need to protect people from poisons while simultaneously protecting them (and their foods) from insects. The insect threat could be measured in precise fiscal terms, but the language of toxicology lacked the sense of immediacy and precision. Thus, Whorton concluded, "The imbalance between these opposed considerations easily tipped entomologists toward optimistic conclusions, and spray residues were generally dismissed as being considerably less dangerous than they actually were."[27] English regulators had been far more aggressive about limiting exposures to arsenic residues, limiting arsenic to 1/100 of a grain per pound (equivalent to 1.43 mg/kg)[28] on fruit in 1903 in direct response to the findings of a Royal Commission on Arsenical Poisoning, which was led by Lord Kelvin.[29] U.S. regulators ignored this standard until 1925, when the English threatened to ban American fruit imports. In 1927, the FDA restricted apples intended for export to 1/100 of a grain per pound of arsenic trioxide (the 1903 British standard), but apples intended for domestic consumption could have as twice as much arsenic per pound.[30]

In 1933, C. N. Myers, a physiologist, and Binford Throne, a clinician, jointly presented the results of several years of research into the risks of

TOXICOLOGY EMERGES IN PUBLIC HEALTH CRISES

spray residues at the eighty-fifth meeting of the American Chemical Society. The researchers acknowledged the considerable costs of international legislation regulating pesticide residues, but they wondered about the costs in terms of human life and health. Previous studies had quantified the arsenic residues on fruit from 0.08 to 0.77 mg per apple and concluded, "No fruit carried sufficient lead arsenate to cause fatal poisoning through the consumption of one piece." But such a statement regarding acute toxicity (resulting in death) neatly sidestepped the problem of chronic toxicity as did subsequent comments, for example, "The case is not clear as to the possible injurious effects from long continued daily consumption of fruits carrying relatively small residue."[31] Although there were cases of acute poisoning in which victims died within days of consuming contaminated fruits or vegetables, Myers and his collaborators worked to sharpen the picture of risks associated with increased arsenic use in a variety of products, among them glucose, drugs, lotions, tobacco, foods, fruits, vegetables, larvicides, and especially insecticides. The risks included eczema, keratosis, peripheral neuritis, disturbance of vision, and neurological symptoms. In addition to the risks borne by humans, Myers noted the destruction of large numbers of bees and birds as a direct result of lead arsenate spraying.[32]

In *100,000,000 Guinea Pigs*, Kallet and Schlink decried spray residues and condemned the FDA for apparent indifference to the problem. The *Guinea Pigs* authors cited arsenic poisonings in the U.S. and England and wondered why the U.S. had been so slow to adopt stricter standards for arsenic residues on fruits and vegetables despite ongoing scientific research and congressional hearings. Of greater concern, however, were lead residues (concern about lead exposures drove the development of occupational medicine, see below). The FDA recognized the considerable hazard posed by lead and banned from the market fruits and vegetables containing residues of lead. However, there was little evidence of enforcement of this strict standard. One researcher tested apples for residues of lead and arsenic and found that none of the forty-five samples were free of either chemical. More troubling, some of the apples carried sixty times as much lead as arsenic trioxide.[33]

It is this sorry state of affairs that returns us to the frustration of Walter Campbell, administrator of the FDA. When the new assistant secretary of agriculture, Rexford Tugwell, queried him on the FDA's failure

TOXICOLOGY EMERGES IN PUBLIC HEALTH CRISES

to ban spray residues, Campbell responded that the department's commitment to the agricultural industry forced the FDA to adopt a lenient policy with respect to growers. The surprisingly receptive Tugwell agreed with Campbell that the 1906 food and drug law needed revision and within hours secured approval from the president for a revision of the act.[34] Tugwell promptly introduced a complete revision of the 1906 law into Congress, but industry representatives universally condemned the bill, particularly a provision that would have required drugs to be licensed. A conservative homeopathic physician named Royal S. Copeland, however, strove to forge a compromise, which would produce a bill with the potential to pass through Congress. The compromise bill passed the Senate but remained stuck in a House committee.

Meanwhile, by 1930 insecticide use in the United States had exploded to unprecedented levels. Farmers sprayed nearly sixty million pounds (27,215,542.2 kg) of the two most popular insecticides (calcium arsenate and lead arsenate) on crops in 1929. Only rarely did spray residues result in cases of acute poisoning and death, but scientists and physicians began to link common ailments like eczema and stomach upset with chronic arsenic and lead toxicity. International regulation of spray residues on fruit imported from the United States prompted review and revision of standards, but levels remained much higher for fruit sold in the U.S.[35] Even Kallet and Schlink's exposé provoked a relatively minor protest. Most Americans remained complacent, assuming that the 1906 law protected them from contaminated food and drugs, and farmers relied on the Insecticide Act of 1910 to keep insecticides free from adulterants. Nevertheless, by 1933 regulators had received executive approval to revise the pure food law. Tragically, an epidemic of poisonings failed to accelerate the legislative process.

During Prohibition following passage of the Eighteenth Amendment, many people sought alternatives to alcohol. Alcoholic extract of ginger had been available since the nineteenth century as a patent medicine and as such it provided a source of ethanol that could be marketed legally. "Jamaica ginger" or "jake" referred to a fluid extract of ginger that was sold widely during Prohibition. The United States Pharmacopoeia (USP), the official body established to monitor patent medicines, attempted to curb abuse of Jamaica ginger by requiring that the content of the extract contain five grams of ginger per one milliliter of solvent,

13

TOXICOLOGY EMERGES IN PUBLIC HEALTH CRISES

which was usually ethanol. But this formulation tasted so bitter that consumers rejected it as non-potable. Pharmacies and roadside stands sold Jamaica ginger, typically in two-ounce bottles, as a remedy for a host of ills. The high alcohol content (up to 80 percent) meant that consumers could buy a two-ounce (56.7 g) bottle of jake for thirty-five cents and mix it with a soft drink, thus creating an inexpensive intoxicating beverage. USDA agents monitored producers by boiling samples of jake and weighing the remaining solids to ascertain that they conformed to the proportion dictated by the USP. Jake manufacturers cut costs by substituting various adulterants, such as castor oil, glycerin, and molasses, for the more expensive ginger solids.[36]

Triorthocresyl phosphate or TOCP appealed to some jake producers as an additive because, unlike other compounds, it would not evaporate away upon analysis by USDA agents (thereby upsetting the ratio of five grams of ginger per one milliliter of solvent). One researcher referred to TOCP as one of the most stable esters used in commercial organic chemistry. It was used in large quantities under various trade names as a liquid plasticizer and in lacquers, leather dopes, and even airplane finishes. Most manufacturers used castor oil as the adulterant to produce a palatable ginger extract that maintained the appropriate ratio of key ingredients. But Harry Gross, president and general manager of Hub Products Corporation, sought alternatives when the price for castor oil climbed at the end of the 1920s. Gross consulted a Boston chemical wholesaler for suggestions of other stable solvents. The wholesaler initially recommended ethylene glycol, which Gross rejected as too volatile, having subjected the resulting Jamaica ginger compound to mock testing. Next, Gross tried diethylene glycol, which produced similar results. Note: Gross rejected ethylene glycol and diethylene glycol on the grounds that the two compounds were too volatile rather than for their toxicity (see below for detailed discussion of diethylene glycol and the Elixir Sulfanilamide tragedy) Finally, Gross settled on Lyndol, which was a mixture of TOCP. When it passed the volatility test, Gross asked the chemical wholesaler about its toxicity. The wholesaler relayed this question to the Celluloid Corporation, which produced Lyndol, and the chemical company noted that it was presumably nontoxic. Gross purchased 135 gallons (511 liters) of Lyndol and began mixing and shipping the new jake.[37]

TOXICOLOGY EMERGES IN PUBLIC HEALTH CRISES

Near the beginning of 1930, physicians and the media began to report a strange illness that was attacking many people across the southern states. By spring the disease had become an epidemic. In Cincinnati, for example, more than four hundred people checked in to the Cincinnati General Hospital with muscular pain, weakness in both the upper and lower extremities, and rather minimal sensory findings. Since most, if not all, of the victims associated the disease with recent consumption of Jamaica ginger, they referred to it as "ginger jake paralysis" or "jake leg." In February 1930, Ephraim Goldfain described a man who had progressive bilateral foot drop, thus becoming the first physician to recognize the disease. That same day, he saw another case and soon developed a list of sixty-five individuals.[38] Over the course of the next few months, thousands of people suffered ginger jake paralysis. So prevalent was the disease across the southern states (especially Tennessee, Oklahoma, Kentucky, and Mississippi) that "jake leg" developed into a significant theme in blues music of the 1930s.[39] Later there were smaller epidemics in Massachusetts and California.

Two researchers with the U.S. Public Health Service, Maurice Smith and Elias Elvolve, isolated the toxic compound as a phenolic compound (triorthocresyl phosphate—TOCP). Moreover, they were able to determine that the poisoned extract of ginger must have originated at a single source based on the fact that it was sold under at least eight different brands in Cincinnati, Ohio, and at least four brands in Johnson City, Tennessee. To reach this conclusion, they needed to develop a toxicological profile for the contaminated samples of ginger jake. Smith and Elvolve first ruled out poisoning by heavy metals (arsenic and lead), which, as we have seen, occurred frequently during the 1920s. The two researchers eventually obtained thirteen samples of the ginger extract. Five of those were almost definitely paralytic. A test for phenols indicated that every sample that contained phenols caused paralysis. To supplement the chemical analysis and correlation with paralytic samples, the researchers administered those samples that tested positive for phenols to rabbits, which exhibited a symptom complex that included muscular tremors, hyperexcitability, and spastic rigidity. General muscular weakness and flaccid paralysis of all the extremities followed. Treated rabbits died of respiratory failure. Smith and Elvolve conducted similar tests with monkeys and dogs only to find that they did not react like the rabbits.

15

TOXICOLOGY EMERGES IN PUBLIC HEALTH CRISES

Further experiments with a solution that included the same elements yielded essentially the same results, but Smith and Elvolve could not explain the specific relation of the phenolic compound to the various neurological effects in ginger jake victims.

In the second report on ginger paralysis, Smith considered the pharmacological action of phenol esters. The comparison of TOCP with several other phenolic esters revealed the Jamaica ginger adulterant to be far more toxic than other similar compounds. The minimum lethal dose for TOCP in rabbits was 100 mg/kg, but as little as 50 mg/kg produced definite symptoms, which occasionally led to death. It was possible to replicate these results in two monkeys by administering the TOCP subcutaneously rather than orally. Pharmacological testing procedures had not been standardized by 1930, so Smith also tested TOCP on calves and chickens, with similar results to those produced in rabbits and monkeys.

Oddly, unlike other phenols or cresyls, the systemic action of TOCP developed slowly. The initial effects of a lethal dose appeared to be limited to the effects of the alcohol in which it was administered. The characteristic group of symptoms developed after an interval of one to several days. The symptoms of TOCP poisoning combined the manifestations of mild strychnine poisoning with some aspects of phenol poisoning. After reproducing symptoms in four experimental species, Smith concluded that TOCP was "capable of producing specific paralysis of the motor nerves of the extremities in certain species of animals and under certain conditions more or less exactly the same as occurred in thousands of human victims traceable to an adulterated fluid extract of ginger."[40] The fact that the pharmacological action of TOCP did not follow any known rule or law inspired Smith to call for further investigations into the pharmacological actions of chemicals.

By summer 1930 estimates of the number of ginger jake paralysis victims had reached as high as twenty thousand. Subsequent estimates significantly increased the number to more than fifty thousand victims.[41] However, since many of the victims were impoverished minorities, researchers suspected underreporting. Despite the extent of illness, the response of the FDA and other federal offices verged on nonexistent. During congressional hearings, a senator criticized the FDA's administrator for the lack of response to ginger paralysis:

16

SENATOR WHEELER: There was only one thing you could do under the law and should do, and that was to seize that product.

MR. CAMPBELL: And we did, promptly.

SENATOR WHEELER: You did, promptly—*two or three months afterwards.*[42]

Kallet and Schlink, in *100,000,000 Guinea Pigs,* cited the case as a prime example of the failing of the 1906 law: by 1932, only three companies had been fined, but only $50 in two cases and $150 in the third. Later, Gross delayed federal prosecution by promising to cooperate and implicate the "real poisoners." Before a trial could proceed, Gross (and his associates) pled guilty in front of a Boston judge, who released him on two years' probation and with a fine of $1,000.

The ginger paralysis epidemic raised a number of questions regarding food and drug legislation, the responsibility of producers, and the role of toxicology. First, the practice of adulterating products with substances to deliberately change the taste or appearance appears to have received at least tacit sanction. If a manufacturer could replace a harmless substance with one of unknown toxicity, consumers were at considerable risk of exposure. In his search for a solvent to replace castor oil, Gross considered ethylene glycol, diethylene glycol, and TOCP. None of these compounds had clear toxicological profiles. In developing the pharmacology of TOCP, the Public Health Service (PHS) scientists first ruled out arsenic and lead, presumably because these were two of the most prevalent poisons that consumers encountered. Tests of samples of ginger jake on rabbits revealed effects similar to those human victims suffered, but monkeys seemed unaffected. Further tests of phenol and several cresols focused on the minimum lethal dose. Compared to similar compounds, TOCP exhibited a much greater toxicity (100 mg/kg and even 50 mg/kg resulted in symptoms and sometimes death). By administering doses to monkeys subcutaneously, researchers produced symptoms comparable to those experienced by humans and rabbits, but the fact that simian system cleared oral doses without toxicity underscored the importance of multiple test species. Chickens and calves also exhibited symptoms. Nevertheless, a clear toxicological profile for TOCP did not facilitate prosecution of litigation against the manufacturers or distributors, who claimed ignorance. Nor did it follow from the ginger paralysis epidemic

TOXICOLOGY EMERGES IN PUBLIC HEALTH CRISES

that manufacturers should be required by federal statute to conduct toxicity tests on compounds before releasing them for public consumption. The fines levied against the companies that poisoned fifty thousand victims struck contemporaries as trivial. Rather than look to the federal government for protection or consideration, jake leg victims turned to music, finding a measure of comfort in the blues.

The best-known instance of public exposure to a lethal compound was the Elixir Sulfanilamide tragedy. The therapeutic properties of sulfanilamide first came to light in 1932, when two German laboratory scientists developed a red dye by combining sulfanilamide to a naphthalene-containing chemical that they called Prontosil. For the next three years, physicians experimented with the new drug for possible applications in their clinics. The real breakthrough came in 1935 when researchers treated streptococci-infected mice with Prontosil with allegedly remarkable therapeutic results. Gerhardt Domagk, who was a researcher for I. G. Farbenindustrie, announced the discovery of Prontosil, but it was largely unheralded due to the limited details of dubiously perfect results. Although cultures of streptococci did not respond to Prontosil in vitro, scientists at the Pasteur Institute in Paris discovered that the mouse systems broke the bond between the naphthalene and sulfanilamide, leaving sulfanilamide to destroy the streptococcus germs. In making this discovery, the Pasteur Institute clarified that it was sulfanilamide that produced therapeutic effects. Moreover, the French results undermined any hopes that I. G. Farbenindustrie may have held that Prontosil could be patented (sulfanilamide's patent of 1909 had expired).[43]

Americans had also participated in early tests on sulfanilamide. The weak bond between naphthalene and sulfanilamide that characterized Prontosil had not met the standards of earlier researchers. During the 1920s, Rockefeller Institute scientists successfully produced a strong bond between sulfanilamide and quinine, but the bond was so strong that sulfanilamide was never released into the system and thus had no effect on disease. This apparent success was ultimately a significant failure, in that it delayed for nearly a decade research on the chemotherapeutic effects of sulfanilamide, until the German researchers utilized the naphthalene combination.[44]

The quest for new chemical therapies following the success of Paul Ehrlich's Salvarsan inspired a campaign to discover new applications and

TOXICOLOGY EMERGES IN PUBLIC HEALTH CRISES

solutions for sulfanilamide.[45] By 1937 more than one hundred firms manufactured proprietary forms of the drug. When the public became aware that sulfanilamide might be a cure for gonorrhea, people began using the medication independently of a doctor's orders, thereby raising the problem of self-dosing. Compounding this problem was the low cost of sulfanilamide on a daily basis. Whereas the most popular patented medicine for treatment of infections like streptococcus cost up to six dollars each day, the cost of sulfanilamide was only thirty cents daily.[46]

The search for chemicals to combine with sulfanilamide in order to create flexible drug-delivery systems led directly to the Elixir Sulfanilamide tragedy. One of the problems with sulfanilamide was the size of the tablets necessary to contain the correct dosage. Because children were most frequently infected by streptococcus, pharmaceutical companies sought a liquid form of the drug. After a few days of research in July 1937, Harold Cole Watkins, the chief chemist and pharmacist at the S. E. Massengill Company, appeared to have solved the problem by mixing sulfanilamide with diethylene glycol, which the Massengill Company had employed successfully in other drugs in extremely small amounts. Ultimately, Watkins determined that forty grains of sulfanilamide per ounce (91.4 g/kg) of diethylene glycol was the optimum proportion, and he prepared a mixture of eighty gallons (302.8 l) of water, sixty gallons (227.1 l) of diethylene glycol, and fifty-eight pounds (26.3 kg) of powdered sulfanilamide, as well as small amounts of the following flavor-enhancing substances: elixir flavor, saccharin, caramel, amaranth solution, and raspberry extract.

After calling the new drug Elixir Sulfanilamide, Watkins sent it to the Massengill laboratory, where it underwent examination for appearance, flavor, and fragrance. Notably, no one tested the new solution for toxicity.[47] In Watkins's view, Elixir Sulfanilamide was exempt from such testing since, he believed, "The glycols, related to glycerine, had been widely used by drug companies, and, Watkins averred, were well known not to be toxic."[48] With quality control completed, Massengill began commercial distribution of Elixir Sulfanilamide on September 4, 1937, a scant two months after Watkins had initiated experiments with diethylene glycol.

As the Massengill Company was distributing Elixir Sulfanilamide around the United States and Canada, the *Journal of the American*

19

TOXICOLOGY EMERGES IN PUBLIC HEALTH CRISES

Medical Association published an editorial by Morris Fishbein, the journal's editor. The editorial opened with a generous appraisal of sulfanilamide in general: "Seldom has any new drug introduced in medical practice aroused the enthusiasm that has developed for sulfanilamide. Much of this enthusiasm is warranted. The drug is truly remarkable, as indicated by startling results reported in the treatment of various infections."[49] Yet Fishbein expressed concern that chemical companies were studying similar and associated preparations in the search for therapeutic agents that they could market as new products superior to sulfanilamide. Fishbein warned the medical profession to proceed with caution regarding new preparations of sulfanilamide. His particular concern was that the therapeutic and toxic properties of new drugs could not be predicted from their chemical formulas: "Many months of investigations of the pharmacology, toxicology, and clinical application of new preparations under carefully controlled conditions are needed to provide evidence of clinical value."[50] There had already been reports of toxic reactions to the self-medication brought about by rumors that sulfanilamide could cure gonorrhea in forty-eight hours. (Fishbein placed considerable blame for such incidents on unscrupulous pharmacists who willingly sold drugs to anyone over the counter.) In a statement that proved to be tragically prophetic, Fishbein concluded, "Sulfanilamide should not be administered in association with other drugs until definite information is available as to toxic effects."[51]

Given the early concern of the American Medical Association (AMA) with the risks of sulfanilamide, it was appropriate that the AMA was one of the first official bodies to learn of the Elixir Sulfanilamide crisis. Two doctors from the Springer Clinic in Tulsa, Oklahoma, sent a telegram detailing clinical and pathologic effects of poisonings in Tulsa to the AMA on October 15. Even allowing for the constraints of communication by telegram and the restrictions of medical reports, the following excerpt seemed particularly stark:

Total of ten cases. Eight dead. One recovered. One critical. Ages from eleven months to twenty-six years. All received Elixir Sulfanilamide in amounts varying from one-half to seven ounces. Characteristic onset with nausea, vomiting, occasional diarrhea, malaise, later pain over kidney region and abdomen. All developed

20

TOXICOLOGY EMERGES IN PUBLIC HEALTH CRISES

anuria within two to five days after beginning medication. Indications for the use of sulfanilamide were varied. Nine cases hospitalized.[52]

It was in fact this report and the cluster of cases it represents that prompted action by the AMA and the FDA.

On October 16, 1937, a representative of the Kansas City Station of the FDA who had been dispatched to Tulsa, Oklahoma, confirmed the doctors' report by telegram to the FDA headquarters, reporting the deaths of nine individuals attributable to a preparation of sulfanilamide. Of these nine, eight of the victims had been children with streptococcic sore throat and one an adult with gonorrhea. All had taken Elixir Sulfanilamide. The FDA immediately sent inspectors to the Massengill Company headquarters in Bristol, Tennessee, and to distribution centers in Kansas City, New York, and San Francisco. Upon receiving news of the deaths, the Massengill Company issued more than a thousand telegrams recalling all outstanding shipments. These messages warned consumers, salesmen, druggists, and doctors not to use Elixir Sulfanilamide and to return all stocks for credit at the manufacturer's expense.[53] While the telegrams conveyed Massengill's desire to collect all outstanding quantities of Elixir Sulfanilamide, they gave no indication of the dangerous character of the product or the emergency that necessitated the recall of the drug. On October 19 the FDA inspector on location in Bristol issued another telegram to all persons known to have received shipments of the drug from Bristol: "Imperative you take up immediately all elixir sulfanilamide you dispensed. Product may be dangerous to life. Return our expense."[54] On the insistence of FDA field agents, similar telegrams were sent from the Massengill branches in Kansas City, San Francisco, and New York, also on October 19 or shortly thereafter.

Despite the return of numerous shipments to the four distribution centers, the FDA faced the considerable task of confiscating all outstanding supplies of Elixir Sulfanilamide. To accomplish this, the FDA diverted most of its field force of inspectors and chemists to review the thousands of order slips in each of the four distribution centers as well as in wholesale and retail drug stores. Complicating matters was the fact that many of the sales of Elixir Sufanilamide were over-the-counter (not by prescription) to unknown individuals. In the most problematic cases,

TOXICOLOGY EMERGES IN PUBLIC HEALTH CRISES

doctors had either no names or fictitious names for individuals who received the drug. Massengill salesmen also proved hard to locate, and at least one went to jail rather than cooperate with the FDA officials. Most doctors and pharmacists contributed willingly to the confiscation campaign, but a few refused to cooperate and even denied the adverse effects of the drug. One South Carolina doctor admitted to inspectors that he had dispensed just under two pints of the elixir to three white patients and two black patients, none of whom died. Further investigation revealed that in fact the doctor administered the elixir to seven patients, of whom four had died: one white man, one white girl, and two black men. The inspector tracked the cause of death to the gravestone of one of the black victims, where a grieving relative had left a small bottle of Elixir Sulfanilamide with the prescription label of the doctor's office still intact.[55]

Notwithstanding the considerable effort of the FDA, as well as state and local food, drug, and health authorities, Elixir Sulfanilamide caused numerous deaths. The tragedy progressed rapidly throughout the Southeast with deaths reported in Mississippi, Alabama, Georgia, South Carolina, and Texas. On November 1, the AMA Chemical Laboratory calculated the total number to be sixty-one. Later that month, the *New York Times* reported that ninety-three individuals had died from Elixir Sulfanilamide, but the official toll stood at seventy-three as a direct result of the elixir and twenty more deaths associated with the drug, according to the report of the secretary of agriculture regarding the sulfanilamide tragedy. Deaths occurred in fifteen states from Ohio to Texas. Most of the deaths were in southeastern states, perhaps due to the proximity of the Massengill headquarters in Bristol, Tennessee. The AMA Chemical Laboratory mapped the epidemiology of the tragedy.[56] Nevertheless, by seizing 228 gallons (863.0 l) of the 240 gallons (908.5 l) manufactured and more than half of the 11 gallons dispensed in prescriptions or over-the-counter sales the FDA surely avoided a much greater disaster.[57]

While the FDA was containing and controlling the immediate crisis, the AMA began a detailed analysis of the chemistry, pharmacology, pathology, and necropsy of Elixir Sulfanilamide. Three scientists at the University of Chicago performed the chemical examination. Their analysis revealed Elixir Sulfanilamide to be a reddish, somewhat viscous liq-

22

TOXICOLOGY EMERGES IN PUBLIC HEALTH CRISES

uid with an aromatic odor resembling raspberry and anise and a sweet taste. The drug resembled glycerin in general physical character. Further analysis yielded the proportions of the major ingredients of the drug:

Diethylene glycol   72 percent
Sulfanilamide   10 percent weight/volume
Water   15.6 percent

Furthermore, the chemists conducted spectrographic examination that failed to reveal the presence of known poisonous substances, such as lead, bismuth, mercury, or arsenic. In essence, the chemical analysis of Elixir Sulfanilamide confirmed the statements of the Massengill Company regarding the composition of the drug.[58]

E. M. K. Geiling, chair and professor in the Department of Pharmacology at the University of Chicago, also conducted toxicity studies of Elixir Sulfanilamide. Geiling determined the toxic agent by carrying out toxicity experiments on rats, rabbits, and dogs using the following substances: pure diethylene glycol, pure sulfanilamide, Elixir Sulfanilamide-Massengill, and "synthetic" elixir of sulfanilamide (produced by the AMA Chemical Laboratory with pure substances in approximately the same proportions as found in the Massengill elixir). Through a series of experiments, Geiling hoped to determine three things:

1. The toxic and lethal doses of each of the substances when given in relatively small doses three times daily. This information seems particularly necessary since we were not able to find any data in the literature on this specific point.
2. Our experiments were further planned with the hope of being able to reproduce in healthy experimental animals, in about the same time, the clinical and pathologic picture as presented by patients who had taken fatal doses of the Elixir of Sulfanilamide-Massengill.
3. Through our experiments we hoped to discern the toxic ingredient in the Massengill elixir.[59]

Such a strategy laid the foundation for future toxicological investigations. Geiling's results left no doubt as to the cause of the deaths. In rats fed via a stomach tube three times daily, the mortality rate rose to

23

TOXICOLOGY EMERGES IN PUBLIC HEALTH CRISES

100 percent after eight or nine doses of two cc's of diethylene glycol, Elixir Sulfanilamide, or synthetic elixir. By contrast, none of the rats treated with sulfanilamide or water died during the experiment. Thus Geiling and his collaborators concluded that diethylene glycol was the toxic agent in Elixir of Sulfanilamide-Massengill since animals treated with the chemical exhibited the same symptoms as animals treated with Elixir Sulfanilamide and a synthetic elixir formulated from the same ingredients in the same proportions.[60]

As further confirmation, they noted that sulfanilamide alone did not prove fatal to rats, rabbits, or dogs, but the drug did cause convulsions in some of the animals, and the researchers raised the possibility that the drug contributed to tissue damage in animals or human beings with impaired renal function. The final element of the conclusion contained significant implications for future policy, specifically premarket testing on animals: "Our experiments emphasize the importance of administering drugs in divided doses to experimental animals when it becomes necessary to know whether or not a drug has cumulative effects. Errors resulting from an oversight of this important pharmacologic principle may be costly to human lives."[61]

Given that drug therapy often required repeated daily doses, Geiling believed that testing new therapies should reflect such procedures if cumulative effects were to be understood. Such a view began to broaden pharmacologists' perspective to include chronic effects. Geiling further clarified his point: "We can confirm the finding of Haag and Ambrose that the ingestion of 15 cc. of diethylene glycol per kilogram in a single dose by stomach tube proves fatal to rats. This figure, however, is no index of the toxic and possible fatal effects of the drug, if administered in small divided doses, especially since neither the fate nor the mechanism of detoxification is known."[62] The earlier research had isolated the lethal dose of diethylene glycol, which is to say, its acute toxicity, but Geiling distinguished this piece of data from the equally important effects of repeated doses, or chronic toxicity. In this sense, the toxicity of diethylene glycol had broader implications for the science and policy of newly introduced drugs.

Finally, Paul R. Cannon, M.D., also from the University of Chicago, conducted a pathological evaluation of rats, dogs, and rabbits that had died following toxic doses of diethylene glycol, Elixir of Sulfanilamide-

Massengill, synthetic elixir, and sulfanilamide alone. Cannon found a remarkable similarity between the pathological effects of a toxic dose of diethylene glycol, the synthetic elixir, or Elixir of Sulfanilamide-Massengill in animals and a lethal dose of Elixir Sulfanilamide in humans.[63] In addition to the research of Geiling and Cannon, Edwin P. Laug and his colleagues at the FDA published the first toxicological benchmark: the LD_{50} (see below).

An appreciation of the impact of the Elixir Sulfanilamide case on the evolution of toxicology in the United States requires a review of the history of the Division of Pharmacology at the FDA. The FDA was established in 1927, when the activities of the Bureau of Chemistry within the USDA became distinct from agricultural chemistry and the work of the department as a whole. Regulators and regulatory chemists were moved to a separate agency called the Food, Drug, and Insecticide Administration (later, the FDA).[64] However, jurisdiction for the Federal Insecticide Act remained under the USDA. This move, noted Christopher Bosso, separated regulatory activities on behalf of consumers from those on behalf of farmers.[65] The first commissioner of the FDA, Walter Campbell was determined to examine the toxicology of lead and arsenic in response to the widespread use of these chemicals as agricultural insecticides. To oversee this project, he appointed Erwin Nelson, a pharmacologist on leave from the University of Michigan, as the acting chief of the new division.[66]

In 1935 Nelson devoted himself to raising pharmacology, hitherto part of the Division of Medicine at the FDA, to division status. To accomplish his goal, Nelson canvassed various universities for experts in critical aspects of toxicology. Several individuals from the original branch of pharmacology made up the core of the new division: Harold Morris, Herman Morris, Howard Lightbody (a biochemist specializing in the study of enzymes), and W. T. McCloskey. Nelson recruited several additional scientists for the new division: Edwin P. Laug, from the University of Pennsylvania; Lloyd C. Miller, a lipid biochemist trained at the University of Rochester; and Herbert Braun, from the University of Wisconsin. Rather than selecting trained pharmacologists, Nelson sought scientists whose specific expertise could contribute to the toxicological analysis of lead and arsenic. Thus he assembled experts in analytical work, enzymology, animal studies, and pathology.[67]

TOXICOLOGY EMERGES IN PUBLIC HEALTH CRISES

Herbert O. Calvery joined the Division of Pharmacology as Nelson's replacement as acting chief. Other scientists joined the new division. Geoffrey Woodard entered the division as a laboratory apprentice, but he played an important role in acute toxicity studies. Harold Morris moved to the National Cancer Institute, where he became widely respected for his research on cancer. O. Garth Fitzhugh, a specialist in the study of chronic toxicity, replaced him in the area of chronic toxicology in 1938. These individuals transformed toxicology from the study of the effect of a single dose on a single animal to the sophisticated statistical analysis of dose response curves necessary to understand the toxic effects of drugs and other chemicals on various animals and humans. Such precision arose as a direct response to the Elixir Sulfanilamide tragedy.[68]

The FDA mobilized its field scientists against Elixir Sulfanilamide and now through the new Division of Pharmacology it could also respond through laboratory analysis. In reviewing the existing literature, including Geiling's research, Edwin Laug and the other FDA toxicologists realized that no one had developed a method for comparing the toxicity of one substance to another. Although Laug and others suspected that diethylene glycol was the cause of the many deaths associated with Elixir Sulfanilamide, they had to develop an approach that would confirm their suspicions statistically. To do this, the FDA toxicologists were assisted by the ground-breaking research of Chester I. Bliss, who was brought to the FDA as a part-time consultant by Herbert O. Calvery. Bliss had studied with R. A. Fisher, the great English biostatistician who developed and promoted the use of statistics in biology.

Bliss had published a seminal paper, "The Calculation of the Dosage-Mortality Curve," in which he demonstrated the sigmoid character of the typical dosage-mortality curve as established by numerous toxicological studies of a large number of organisms by many biologists. Acknowledging his debt to R. A. Fisher and the trends in biostatistics Fisher inspired, Bliss selected his procedures on the basis of their statistical accuracy and efficiency. Bliss intended to present his techniques for calculating the transformed dosage-mortality curve in sufficient detail so that biologists with limited knowledge of statistics could use them. The value of the dosage-mortality curve to biologists in general, and toxicologists in particular, arose from its ability to describe the variation

26

in susceptibility between individuals of a population. In line with the theory that any given population would show a range of susceptibilities to any given toxin, Bliss demonstrated that mortality could be plotted against dose in a way that would indicate what percentage of any given population would be killed by a particular dose. This experimental technique could determine the precise minimum lethal dose for each organism, since a dosage above minimum effectively killed susceptible individuals. Bliss recommended exposing a series of sample groups of organisms to graduated doses and recording the percentage killed for each sample group. When Bliss plotted this data, it resulted in a sigmoid curve (exactly as predicted). Thus he concluded: "The sigmoid dosage-mortality curve, secured so commonly in toxicity tests upon multicellular organisms, is interpreted as a cumulative normal frequency distribution of the variation among the individuals of a population in their susceptibility to a toxic agent, which susceptibility is inversely proportional to the logarithm of the dose applied."[69] Bliss's explanation and technique were enhanced by a note from Fisher. It examined the case in which there were few or no survivors and appended a method to address this problem.

Bliss's paper appeared in 1935, but the FDA pharmacologists were already familiar with his work, since he worked in their laboratory. The dosage-mortality curve provided an excellent method to evaluate the toxicity of a given chemical. Laug and his colleagues in the Division of Pharmacology were among the first to apply Bliss's method to an actual case when they evaluated the toxicology of glycols and derivatives in response to the Elixir Sulfanilamide disaster. Prior to this study, toxicologists had estimated the doses that were lethal to 10 percent, 50 percent, and 90 percent of a given population, but the results were unsatisfactory, particularly when toxicologists used such data to interpolate a sigmoid curve through a number of points. Laug and the FDA pharmacologists used Bliss's method to calculate and plot the dose-mortality curve for rats, mice, and guinea pigs. They determined that the most useful parameter was the lethal dose for 50 percent of the population (LD_{50}), as this value required the fewest animals for calculation (a minimum of 10). In contrast, arriving at the same level of precision for 99 percent deaths required at least 103 animals. Moreover, Laug and his team calculated standard errors for the LD_{50} and found they could determine the dosage at which 50 percent of the population would be killed in nineteen out of twenty experiments.

Citing the studies of Geiling and others at the University of Chicago, Laug reflected on the importance and probable significance of his team's findings for humans. Although the FDA researchers had confirmed the findings of Geiling and others, Laug advised restraint in applying the results to humans: "It seems proper at this point to reemphasize the inadvisability of attempts to interpret experimental data on laboratory animals directly in terms of man. It is entirely too dangerous. The present investigation confirms again the wide variations that may occur between species."[70]

Although they presented the derivation of LD_{50} as a valuable new tool in toxicology, the FDA researchers acknowledged that neither acute nor chronic toxicity gave a complete picture and scientists should assess both dimensions of toxicology. They argued for establishing the most complete toxicological profile possible. Ideally, such a profile would include both acute and chronic toxicity studies as well as pathological evaluations. In addition, certain glycols produced acute effects at high doses without producing chronic effects at lower doses. Laug contrasted this toxicological behavior with that of the well-studied heavy metals and narcotics: "Again, although small doses of lead, mercury, selenium, fluorine, narcotics, etc., are acutely toxic, it is the insidious character of their chronic toxic effects that makes them even more dangerous."[71] As of the 1930s, the study of chronic toxicity had not progressed as far as the study of acute toxicity, except in the case of heavy metals and certain drugs.[72] Nevertheless, the model of chronic toxicity provided by heavy metals was hardly universal, as Laug noted in the statement above.

Methods for deriving dose-response curves and LD_{50}s were certainly among the most important legacies of the Elixir Sulfanilamide tragedy. Not only did the approach transform regulatory science as it was conducted at the FDA, industry adopted the same procedures for its analyses of new chemicals (at the FDA's direction). Long after he retired, Laug would recall the significance of the division's initial exploration of LD_{50} as one of the first successful applications of statistical approach to toxicology: "I think it was the most significant thing that we did . . . in those days there was not much precision when determining toxicity.

And what we did was by the use of statistics, we made it possible that when you treated animals with something toxic, you could create a curve, a slope, and the significance of that was that you could then compare it to something else and that was the point, LD_{50}."[73] A small group of FDA toxicologists in the Division of Pharmacology transformed the field of toxicology. More than any other single procedure, LD_{50} became the benchmark in most initial studies of toxicity for pharmaceuticals and environmental chemicals.

Still, the question remains: Why did the Division of Pharmacology, which was established to evaluate the hazards of lead and arsenic in insecticides, develop the LD_{50} in response to the Elixir Sulfanilamide disaster? The Division of Pharmacology lost its funds to investigate lead and arsenic in 1937 or 1938. According to Geoffrey Woodard, another FDA pharmacologist, "Well, the insecticide problem was the original basis for having set up a division. Now as Ed [Laug] said, it was a political football and I remember we worked through '37 or '38 until New Year's Eve—right up until midnight on New Year's. Because Congress cut off our funds and said there was to be no more work on lead and arsenic."[74] In Woodard's view, which was shared by the other FDA pharmacologists, lead and arsenic insecticides had become politically charged.[75] In response to increased scrutiny, apple growers had appealed to their congressmen, who voted to transfer the authority for the examination of insecticides to the Public Health Service (PHS) at the National Institutes of Health (NIH) by revising the appropriation act of the FDA so that none of the funds could be used for the study of toxicity of lead and arsenic. The restriction of funds did not stop Calvery and Laug from publishing on the risks of lead compounds, however.[76] Deprived of its work on the toxicology of lead and arsenic insecticides, the Division of Pharmacology concentrated its considerable effort on the study of Elixir Sulfanilamide, the glycols, and eventually other chemicals. Christopher Bosso revealed that shifting research on insecticides from the FDA to the PHS meant a move away from long-term laboratory studies based on experiments with laboratory animals to extrapolate chronic effects. In contrast, the PHS emphasized field surveys that questioned farmers about their health, which could reveal acutely toxic effects, but for the FDA, the PHS approach was inadequate for determining longer-term chronic effects.[77]

TOXICOLOGY EMERGES IN PUBLIC HEALTH CRISES

While the FDA strove to eliminate the risks posed by Elixir Sulfanil-amide, and the AMA with the assistance of Geiling and other University of Chicago faculty conducted the chemical and toxicological analysis of the elixir, Massengill attempted to defend its product. The company had earlier been convicted and paid fines for violations of the Food and Drugs Act, in September 1934 and March 1937. H. C. Watkins, Massengill's chief chemist, had been cited by the solicitor of the Post Office for distributing a medicine alleged to reduce weight, to bring about "perfect slenderness," and to cause the body to acquire "a trim, youthful, athletic look." To avoid charges of fraud, Watkins filed a stipulation agreeing that the sale of the product would be abandoned and not resumed at any future time.[78] Unfortunately, under the PFDA, the only basis for action against the interstate distribution of the "elixir" was the allegation that the word implied an alcoholic solution, whereas the product was a glycol solution. It was for this reason that S. E. Massengill, owner of the eponymous company, could claim in a letter to the AMA, "I have violated no law." Although Massengill's statement conformed to the letter of the law, regulatory bodies generally believed that most drug manufacturers recognized a greater responsibility to the public. Such ethical obligations did not concern Massengill. He refused to take responsibility for the deaths, instead blaming them on the "bad effects" of sulfanilamide.[79] Massengill's statement betrayed his complete lack of knowledge regarding the toxicity of Elixir Sulfanilamide. Even in the absence of controlled animal experiments, like those conducted at the University of Chicago and the FDA, a simple literature review would have revealed a careful analysis of ethylene glycol (the active ingredient in antifreeze) and some of its derivatives, including diethylene glycol.

Specifically, Massengill and Watkins would have discovered a paper written by Oettingen, one of the fathers of pharmacology in America. Oettingen, who at the time taught pharmacology at Western Reserve University in Cleveland (and later served as the director of the Haskell Laboratory), tested the toxicity of the glycols on various animals, including rats and frogs.[80] Regarding the therapeutic potential of diethylene glycol and ethylene glycol, Oettingen wrote: "Ethylene glycol and diethylene glycol may be of interest for therapeutic use, as solvent and as vehicle. With these substances the local irritation is comparatively small. Their application to the skin seems to be without risk. Given orally in

30

larger doses, they may produce severe gastro enteritis and systemic symptoms."[81] Still, even if they had completed a literature review, Massengill and Watkins might have argued that the benefit of the drug sulfanilamide outweighed the unknown risk of gastro enteritis. It is also possible that they had experience with one of the nontoxic glycols, such as polyethylene glycol, which is nontoxic and an effective laxative (widely marketed today as an over-the-counter therapy).

Given their academic and professional experience, Massengill and Watkins should have been qualified to appreciate the inherent risks associated with their enterprise. Yet FDA investigators found Watkins to be particularly unconcerned about the hazardous effects of his products: "What impressed [Theodore G.] Klumpp and [William T.] Ford most about Watkins was what they deemed a certain callousness in his conversation. He spoke of a preparation of colloidal sulfur he had devised. When marketed, this compound resulted in the death of a number of people. 'Mr. Watkins told about this event,' Klumpp wrote, 'as if it were an ordinary incident in the business of making and marketing pharmaceuticals.' "[82] Indeed, Watkins seemed to doubt the possibility of toxic effects altogether. In his interviews with FDA officials, he claimed that when he first heard reports the Elixir Sulfanilamide had been linked to numerous deaths, he had personally taken a huge oral dose of diethylene glycol without ill effects. FDA officials doubted this story and dismissed its claim as a "futile heroic gesture," which could not make up for Watkins's failure to test the drug properly before its distribution.[83]

Although generally regarded as inadequate, the PFDA was the only relevant legislation available to FDA in bringing charges against the S. E. Massengill Company in the Elixir Sulfanilamide case. As owner of the company, S. E. Massengill was particularly susceptible to the sanctions of this act because the two earlier prosecution charges exposed the company to more severe penalties as a second offender. For his part, Massengill continued to proclaim his innocence, arguing that Elixir Sulfanilamide had helped more people than it had harmed. The company did fire Watkins, and it admitted a level of moral responsibility when it settled numerous suits brought by the families of those who had died as a result of the elixir.[84] Nevertheless, the FDA mounted a strong case against the company, confirming that diethylene glycol was the agent of toxicity in the elixir and assembling a team of experts to testify to the toxicity of the drug. These included Perrin H. Long of Johns Hopkins, who later wrote a monograph on the clinical and experimental use of sulfanilamide, as well as Geiling and Cannon of the University of Chicago. In addition, FDA scientists prepared testimony and exhibits for the trial. On the first day of the scheduled trial (October 3, 1938), Massengill appeared and revised his plea to guilty, and he received a fine of $150 on each of 174 counts amounting to a total of $26,100, the largest fine ever levied under the provisions of the 1906 act. Adding to the tragedy, Harold Watkins avoided further condemnation by committing suicide before the trial. Nevertheless, it was the calamity of Elixir Sulfanilamide that finally motivated Congress to pass new legislation regarding food and drugs.

As the supervising agent, Henry A. Wallace, the secretary of agriculture, completed an extensive review and analysis of the deaths attributed to Elixir Sulfanilamide. The report responded to two resolutions in the United States Congress: House Resolution 352 of November 18, 1937, and Senate Resolution 194 of November 18, 1937. These resolutions mandated an investigation into the Elixir Sulfanilamide debacle and its implications for food and drug legislation. After thoroughly reviewing the case, Wallace developed four broad recommendations for legislation. First, he recommended the licensed control of new drugs to insure that they would not be generally distributed until experimental and clinical tests showed them to be safe for use. As a corollary to this recommendation, the secretary defined exactly what constituted a "new drug." In order to justify drug licensure, he noted the importance of safety: "In the interest of safety, society has required that physicians be licensed to practice the healing art. Pharmacists are licensed to compound and dispense drugs. Electricians, plumbers and steam engineers pursue their respective trades under license. But there is no such control to prevent incompetent drug manufacturers from marketing any kind of lethal potion."[85] Another provision of the recommendations called for the prohibition of drugs that were dangerous to health when administered in accordance with the manufacturer's directions. As self-evident as this stipulation seems, recall that the only grounds on which to prosecute S. E. Massengill Company was the minor issue that the drug was not technically an elixir as it had not been combined with alcohol. The secretary also stipulated that drug labels bear appropriate directions for use

TOXICOLOGY EMERGES IN PUBLIC HEALTH CRISES

and warnings against probable misuse. Finally, he called for the prohibition of secret remedies by requiring that labels disclose fully the composition of drugs. The new language would broaden the legislative authority of the FDA, enabling the prosecution of manufacturers of dangerous drugs then on the market. Under the Pure Food and Drug Act of 1906, the FDA could not bring even the most trivial of charges against many of the dangerous drugs. In a recent analysis of the role of the tragedy in food and drug legislation, the government scholar Daniel Carpenter indicated the significance of the efforts of Wallace and FDA Chief Campbell, "Yet it is no understatement to say that Campbell and Wallace's document forms the originary basis of modern pharmaceutical regulation the United States and much of the industrialized world."[86]

As a result of the Elixir Sulfanilamide disaster, pharmacologists developed new standards for the approval of drugs for public use. Of the many scientists who analyzed the toxicity of Elixir Sulfanilamide, one of the most outspoken was E. M. K. Geiling. It was Geiling's conviction that drugs needed to undergo a thorough toxicological examination before release. Based on his work with Elixir Sulfanilamide, Geiling developed a nine-part analysis utilizing animal experiments for new drugs. Geiling's proposed method required knowledge of the exact composition (qualitative and quantitative) or the detailed method of preparation of the product. Repeatedly, scientists and legislators cited lack of knowledge of the contents of Elixir Sulfanilamide and its potential effects as a serious flaw in the PFDA. The new drug analysis demanded acute and chronic toxicity studies of animals of different species at varying dosage levels given that studies of a single species at constant dosage levels could be misleading. During the course of chronic toxicity studies, Geiling encouraged careful and frequent observations of the animals in order to establish a composite picture of the clinical course of the drug. Geiling noted that the data on many drugs were very deficient in this respect.

Drug analysis also called for careful pathologic examination of the tissues with appropriate stains. Animal experiments would also facilitate the study of the effects of experimental lesions of various important excretory or detoxifying organs, particularly the kidneys and liver, on the action of the drug. Related to this requirement is the determination of the rate of absorption and elimination of the drug, its path and manner of excretion, and the concentration levels in the blood and tissues at

33

TOXICOLOGY EMERGES IN PUBLIC HEALTH CRISES

varying times after administration. Another strategy that Geiling recommended was the analysis of potential interactions between the drug
and foods or other drugs (as an example, Geiling offered the contraindi-
cation of sulfanilamide and magnesium sulfide). Finally, Geiling called
for careful examination for idiosyncrasies or untoward reactions of new
drugs.[87]

With this method of analysis, Geiling defined the toxicological assessment of drugs, but he acknowledged that his expectations probably
exceeded the dedication of pharmaceutical chemists: "It is recognized
that some will consider these safeguards to be too rigid and that they
may simply be considered an ideal. It can correctly be charged, in fact,
that some of the pharmacopeia drugs have not been studied along such
lines."[88] Geiling intended, however, to elevate the standards of drug approval to new levels and thereby avoid other disasters: "Admitting this,
it is nevertheless regrettably true that many human lives have been sacrificed by the failure to meet the standards of these preliminary tests and
that many more lives will be sacrificed if such standards are not put into
effect. Any essential compromise with these requirements will inevitably
exact a toll of deaths or injuries among the public. The life and safety of
the individual should not be subordinated to the competitive system of
drug exploitation."[89] Geiling believed that such standards would protect
the public from untested drugs with unknown effects. Finally, Geiling
hoped that new standards would direct drug manufacturers toward the
development of new and genuinely valuable agents in the treatment of
disease in place of spending enormous sums on advertisements.[90]

Industry representatives contradicted Geiling's notion of the role of
advertising in drug manufacture and public health. A former president
of the National Association of Insecticide and Disinfectant Manufacturers, Lee H. Bristol (vice president of the Bristol-Meyers Company)
claimed that manufacturers of proprietary drug products favored drastic
control over untried potentially dangerous drugs. He also noted that
the sulfanilamide elixir was not a consumer-proprietary product intended
for self-medication, but one prescribed by physicians. The responsibility
for the calamity rested with the drug manufacturer who used an inade-
quately tested and dangerous solvent for a potent drug and the physician
who prescribed or used the elixir without being familiar with its therapeutic effects. The *New York Times* reported an interesting inversion of

34

Geiling's statements cited above: Bristol argued that "advertising was the greatest safety measure ever devised for the protection of the consumer" and pointed out that Elixir Sulfanilamide was an unadvertised product belonging to a group of so-called ethical specialties.[91] Bristol went on to concede that government should have control over such potent drugs as sulfanilamide elixir and said that no reputable manufacturer of packaged medicine stands in the way or opposes legislation to protect the public from such calamities. He added the following qualification to his statements, however, "Experienced and responsible manufacturers of carefully compounded products should not be penalized simply because their products are advertised to the public."[92]

Bristol's comments were particularly significant in light of his association with the National Association of Insecticide and Disinfectant Manufacturers for his statements applied equally well to insecticides and drugs. The difference between Geiling's perspective and Bristol's was one of emphasis. Whereas Geiling believed that advertising drugs necessarily detracted from meticulous evaluation of the toxicology and pharmacology of new drugs, Bristol argued that advertising served to inform consumers as to the qualities of new pharmaceutical agents. Neither position was wholly sustainable independent of a deeper understanding of how pharmaceutical companies conducted their research, development, production, and advertising. Geiling and Bristol both clearly recognized that the Elixir Sulfanilamide case revealed an urgent need for change in food and drug law but differed as to the precise form of such legislation.

Other calls for new legislation were more explicit. In an editorial for *Hygeia*, a family health journal, Fishbein argued that the PFDA of 1906 was passed at a time when modern advertising was in its infancy and that more food and drugs were sold in 1937 by advertising than by claims made in other ways. More troubling was *Hygeia*'s contention that present food and drugs law provided no potent weapon against false and fraudulent advertising of foods, drugs, devices, or cosmetics. At its core, the 1906 Act did not provide for standards of purity, potency, wholesomeness, or labeling of foods and drugs. Geiling added toxicity standards to this list. *Hygeia* also decried the many loopholes contained in existing legislation.[93] More than any other incident, the Elixir Sulfanilamide disaster crystallized ideas regarding food and drug legislation.

TOXICOLOGY EMERGES IN PUBLIC HEALTH CRISES

To this point in the story, we have seen how several health tragedies, particularly the Elixir Sulfanilamide disaster, transformed toxicology. Moreover, we have seen how researchers at universities, government agencies, and industry refined the toxicological approach. But how did such developments influence food and drug law? The journalist Philip J. Hilts argued that there were two basic requirements for novel legislation: "A bill must *already be present* in Congress, and legislators and significant elements of the public must already be *educated and paying attention* when the crisis hits."[94] He added the corollary that the crisis must also involve children. When the Elixir Sulfanilamide disaster first appeared in newspapers, and *100,000,000 Guinea Pigs* reached bookstores and libraries, Congress was reviewing the Pure Food and Drug Act of 1906, which regulators and the public regarded as inadequate protection for consumers. Republican control of Congress during the 1920s was not conducive to legislative reform, but the arrival of Franklin Delano Roosevelt and the New Deal promised change. As we have seen, when the story of the Elixir Sulfanilamide tragedy broke in 1937, a compromise bill combining elements of the Tugwell bill and the Copeland bill had passed the Senate but remained stalled in a House committee.

It was the sulfanilamide tragedy that prompted sharp demands for more effective legislation from Campbell, Wallace, and Fishbein of the AMA. Reinforcing these demands were those of the media and the public. Congress reopened its consideration of revising the PFDA. Although drug licensing authority and continuous inspection of drug manufacturing did not become part of the new law as the FDA had hoped, Congress did include a "new drugs" provision: Section 505. This provision clearly defined a new drug and forbade the sale of new drugs in interstate commerce before their manufacturers had demonstrated to the FDA that they were safe. In their application to the FDA, manufacturers had to include samples of the drug, a description of the methods of manufacture, and full reports of investigations made to determine whether or not the drug was safe for use. In the absence of protest by the FDA within a specified amount of time, the drug could be released for interstate sale. Alternatively, the FDA reserved the right to refuse an application, effectively barring a drug from interstate commerce pending appeal. Thus the Federal Food, Drugs, and Cosmetics Act of 1938 (FFDCA) significantly promoted the regulatory authority of the FDA,

36

albeit still falling short of licensing.[95] In an insightful analysis of the bureaucratic history of the FDA's efforts toward revision of food and drug legislation, the historian of medicine Gwen Kay concluded: "Ultimately, one tragedy did what five years of public relations by the FDA could not do, which was to convince members of the House of Representatives that revision of the 1906 act was important."[96]

Although arsenic and lead spray residues and the ginger jake paralysis epidemic affected thousands of lives and caught the attention of regulators and consumer advocates, it was the Elixir Sulfanilamide tragedy that ultimately set the stage for the development of a scientific understanding of environmental risk as posed by pesticides. Discontinuation of congressional appropriations restricted the FDA from the study of lead and arsenic insecticides, but this shift proved to be inspirational. The Division of Pharmacology at the FDA developed the LD_{50} in order to determine the toxicity of glycols and derivatives. Meanwhile, at the University of Chicago, E. M. K. Geiling developed a broad-based approach to the toxicology of new and unfamiliar chemicals. The Division of Pharmacology at the FDA and the Department of Pharmacology at the University of Chicago served as centers for toxicological research during World War II. The research programs at the two institutions diverged after the Elixir Sulfanilamide tragedy, but they would converge in time. In addition, the study of industrial disease had developed as researchers studied workers in lead factories. This research led to the establishment of in-house toxicology laboratories at DuPont, Dow, and Union Carbide. Meanwhile, partially in response to consumer advocacy and partially in response to the Elixir Sulfanilamide tragedy, federal legislators significantly revised the PFDA to produce the Food, Drugs, and Cosmetics Act of 1938, which would define pesticides regulation for several decades. The FFDCA expanded the FDA's authority in regulating insecticides but failed to provide a provision for licensing. During and after World War II, the Division of Pharmacology at the FDA applied the new techniques of toxicology to DDT, a new insecticide introduced for the control of insect-borne disease in World War II.

CHAPTER 2

DDT and Environmental

Toxicology

After the conclusion of the World War II, *Time* magazine attributed the Allied victory to two new technologies (among others): the atomic bomb and DDT.[1] Years later, Rachel Carson would compare DDT to the effects of atomic radiation in a less favorable light. Early tests of the toxicity of DDT were conducted during the war, and initial reports indicated that the compound was one of the most effective insecticides ever created. In addition, results from Naples, Italy, and islands in the South Pacific indicated that human toxicity was very low. Was DDT the magic bullet that economic entomologists had sought for so long? How effective was DDT on the many insects that threatened crops in the United States and elsewhere? What acute and chronic effects might this new chemical pose for human, animals, and other plants?

During and immediately after World War II, DDT (dichloro-diphenyl-trichloroethane) underwent extensive scrutiny by a range of scientists, including economic entomologists; public health officials; and wildlife biologists from such organizations as the USDA, PHS, FDA, the U.S. Fish and Wildlife Service (FWS), research universities, agrochemical companies, and private institutions. Moreover, scientists' experience with DDT established modes of analysis for assessing other synthetic insecticides and chemicals. Indeed, scientific scrutiny of DDT transformed toxicology as scientists struggled to determine the toxicity of the new chemical to target insects, including most economically significant species—bedbugs,

38

lice, cockroaches, mosquitoes, fleas, flies—and many others, but economic entomologists also needed to determine the impact of DDT on beneficial insects, for example, on honeybees.

In order to represent the range of early toxicological analysis of DDT, I have examined many of the early investigations of target insects, laboratory animals, wildlife, and humans. Because DDT and other chlorinated hydrocarbons (the class of synthetic compounds to which DDT belongs) had been developed as insecticides just before the U.S. entered World War II, scientists had little experience with the various forms (emulsions, dusts, solutions), doses (1 percent, 5 percent, 10 percent), or concentrations (5 pounds/acre, 1 pound/acre, 0.1 pound/acre) and did not know which would be the most effective in controlling insects without posing risks to other organisms, including plants. Since all of these factors needed to be determined, most studies tested various forms, doses, and concentrations.

In examining these studies, I have sought to capture the complexity of the experiments as well as the considerable variation in the type of DDT employed. Because of the inherent uncertainties in the forms of DDT just noted, and the variety of organisms involved, there was very little continuity linking one study to another. Different organizations were involved in the analysis of DDT, and some had competing agendas, which further complicated matters. Although these factors render the following analysis of preliminary studies of DDT somewhat disconnected, in the aggregate the studies marked a crucial episode in the history of toxicology and environmental risk.

DDT was initially synthesized by an Austrian chemistry student—Othmar Zeidler—in 1873. Zeidler was interested only in the process of synthesizing DDT, not in its potential uses. Not until 1939 did Paul Müller, a staff scientist at the Geigy firm in Basel, Switzerland, discover that DDT was extraordinarily powerful at killing insects on contact with long-lasting effects after application.[2] As historian John H. Perkins has noted, seven criteria for distinguishing a "desirable commercial product" guided Müller's research:

1. Great toxicity toward insects.
2. Rapid action, that is, onset of the paralysis within a few minutes.
3. Zero or weak toxicity toward warm-blooded animals, as well as toward fish and plants.

4. Absence of irritating action to warm-blooded animals, and little or no unpleasant smell.
5. Polyvalent action extending to the greatest possible number of insects.
6. Long duration of action, that is, great chemical stability.
7. Low price = economic advantage.[3]

Müller found that DDT exhibited all of the qualities with the exception of the second, "rapid action." It could take hours for the insect to feel the lethal effects of the new insecticide. In a comparison with other insecticides, including nicotine, rotenone, pyrethrum, thiocyanates, and phenothiazines, Müller discovered that only DDT met the criteria of long duration of action. DDT also had considerable economic advantage over two of its primary competitors, rotenone and pyrethrum.[4] The toxicity of the substance to animals and fish was a subject for further investigation. Geigy circulated information regarding an unspecified "new insecticide" to its subsidiary firms within two years of Müller's discovery, but the U.S. subsidiary ignored the information. But after Geigy sent samples of DDT to the subsidiary, it recognized its considerable potential and forwarded samples to the USDA, which had been investigating new insecticides since 1940 as part of the war effort, under the supervision of Percy Nichol Annand, who had become chief of the Bureau of Entomology and Plant Quarantine (BEPQ) late in 1941.[5]

The BEPQ, under the auspices of the USDA, conducted the first tests of DDT in America.[6] Founded in 1872 to develop control methods for boll weevil and gypsy moth populations, the BEPQ originally explored biological control methods. Under Leland Ossian Howard's directorship (1894–1927), the BEPQ shifted to chemical control methods using such well-known insecticides as lead arsenate, Paris green, sulfur, and kerosene.[7] The bureau's laboratory in Orlando, Florida, directed by Edward F. Knipling, conducted many of the initial tests of efficacy and toxicity on DDT on the behalf of the U.S. armed forces. Knipling (1909–2000) was a young economic entomologist who had held various posts with the USDA in Iowa, Georgia, Texas, and Oregon, before he became director of the Orlando laboratory. Under Knipling's direction, the Orlando laboratory tested the toxicity of DDT against several kinds of lice, including body louse (*Pediculus humanus corporis*), crab louse, and head louse in December 1942. By then DDT research had reached new levels of urgency as conditions in Naples, Italy, had reached a state of emergency as typhus threatened the city. Typhus, a deadly disease that thrives in crowded conditions, was carried by body lice.

Raymond C. Bushland (1910–1995) and his staff from the Sanitary Corps of the U.S. Army developed methods for testing various insecticides including DDT on body lice. First, they performed simple tests on approximately 7,500 chemicals in which they exposed lice to chemicals on treated cloths placed in small beakers. Based on the results, the BEPQ scientists eliminated all but a few chemicals from consideration. In the second procedure, utilizing the reduced set of chemicals, they placed the chemically treated cloths on the arms and legs of fully informed volunteers. In the final experiment, volunteers dressed as soldiers, wearing wool winter underwear that was treated with a promising chemical in powdered form. The researchers placed several hundred lice in various stages of maturity and over a thousand eggs *directly on the volunteers' clothing.* One challenge of this research was maintaining colonies of thousands of lice requiring regular feedings. Also, volunteers had to endure the discomfort of feeding up to fifty thousand body lice once every two weeks. These tests revealed the value of two chemicals: MYL louse powder (made from pyrethrum) and DDT louse powder.

Bushland noted that DDT was effective in louse powder and proved to be "highly effective and longer lasting than any other louse treatment known to be in use."[8] DDT was similarly effective when impregnated in clothing. It remained effective for three to five weeks when clothing was not washed, and once-a-week washings reduced effectiveness to two or three weeks. DDT also proved effective against head and crab lice, particularly in a liquid preparation. After the FDA certified DDT powder as safe for humans, the BEPQ recommended it to the armed forces as a safe and effective louse powder. Historian David Kinkela has shown that efforts at the Orlando laboratory were augmented by a lab at the Rockefeller Foundation in New York. Between 1942 and 1944, Rockefeller scientists tested DDT on a range of subjects, including medical students in New York, conscientious objectors in New Hampshire, and an unwitting civilian population in Mexico.[9] The extensive use of DDT by the armed forces (in cooperation with the Rockefeller Foundation) helped to prevent a potential typhus epidemic.[10] Several years later, Knipling

DDT AND ENVIRONMENTAL TOXICOLOGY

predicted even greater achievements: "We now believe that DDT provides the means of not only controlling lice and typhus but of eventually
eradicating typhus from the earth."[11]

Early testing at the Orlando laboratory demonstrated the efficacy of
DDT against numerous other insects.[12] Repeatedly, DDT's property of
persistence, lacking in most if not all of the other synthetic insecticides,
proved to be important in insect control. For example, bedbugs posed a
challenge to economic entomologists because both the common species
(*Cimex lectularius*) and a tropical species (*C. hemipterus*) remained con-
cealed and protected when insecticides were sprayed in a building. Entomologists at the Orlando laboratory found that DDT in various solutions
of kerosene produced 100 percent mortality for bedbugs at least two
months after application, and the strongest solutions completely eliminated them for more than nine months.[13] Likewise, DDT spray controlled
ticks both in the wild and on dogs. Researchers also noted that none of the
experimental dogs showed any ill effects from the treatment.[14]

Even roaches could not withstand the effects of DDT. Researchers applied various spray and powder solutions of DDT to army mess halls that
were heavily infested with German cockroaches (*Blattella germanica*).
Within thirty minutes of spraying, 2,000 cockroaches lay dead and dying
on the floor of one building. After twenty-four hours, scientists counted
more than 4,000 nymphs (juveniles) and adults on the floor of each of
three buildings. This study suggested that the virtually invisible residue
was sufficiently toxic to protect buildings from German cockroaches for
just a few days.[15]

Laboratory tests of the toxicity of DDT to twenty different insect
pests, including blister beetle, Colorado potato beetle, imported cabbage
worm, and several weevils, aphids, and termites, yielded 100 percent mortality in most cases using either a spray (1 pound per 100 gallons of water)
or a 3 percent dust of DDT. Only termites evaded these deadly effects.
Yet DDT seemed to repel the termites, which desiccated on the surface
rather than passing through the DDT and sand mixture.[16]

Like termites, aphids proved resistant to DDT. Scientists recorded
aphid mortality of a mere 5 percent with DDT, whereas nicotine destroyed 86 percent of the aphids. Dusting wet plants with DDT did not
injure the plants (one of the problems associated with heavy metal insecticides).[17] Other insects, including the cotton leafworm (*Alabama argil-*

42

acea) and the boll weevil (*Anthonomus grandis*), thrived in spite of DDT.[18] One particularly refined study demonstrated that DDT could encourage population growth in certain insect species. In DDT-treated plots, the yellow sugar cane aphid population exploded to six times that of plots left untreated or plots treated with synthetic cryolite. The entomologists concluded, "It is apparent that the great increase in aphids resulting from the use of DDT would be a serious drawback to its use in control of the sugarcane borer."[19]

In addition to nuisance insects like flies, ticks, bedbugs, and cockroaches, economic entomologists sought to determine the efficacy of DDT on the insects that threatened important economic crops, such as cotton and tobacco. To that end, they tested DDT on numerous cotton insects, but it proved to be far less effective against these insects than nicotine.[20] Many scientists noted that in the field, the fruit and foliage of peaches, plums, or grapes were apparently not injured by DDT use. In a comparative study of insecticides, a team of scientists noted the advantages of DDT over other insecticides: "It remains effective for a longer period than derris and has the advantage over the lime and the lead arsenate sprays in that it leaves no conspicuous residue on the fruit and foliage."[21]

Despite the lack of any apparent effect on plants, several researchers found ancillary effects on nontarget insects. Like aphids and cotton insects, the European red mite (*Paratetranychus pilosus*) became noticeably more abundant on trees sprayed with DDT. Researchers noted that lady beetles were absent from the infested plots, and their return brought the mites under control. Of course, DDT destroyed lady beetles. Similarly, large populations of the common red spider (*Tetranychus* spp.) ran rampant on the apples and under the bark of all trees sprayed with DDT, though red spiders were difficult to find in other plots.[22] A few scientists expressed concern over the possible threat of DDT to beneficial organisms. One clearly deleterious effect was the destruction of lady beetles and the ensuing proliferation of cotton aphids. Worried about the impact on honeybees, they discovered that DDT, indeed, poisoned honeybees at 1 percent concentration and even lower.[23]

DDT appeared to be a wonder insecticide, able to control a wide range of insects (i.e., it was a broad spectrum insecticide). Underneath the shouts of praise, however, there were murmurs of concern. DDT did not seem to affect larger organisms or plants, but certain invertebrates survived its effects and proliferated as DDT destroyed their predators. Equally disturbing was the possibility that certain insects developed resistance to DDT's effects. The toxicity of DDT to the Mexican fruit fly (*Anastrepha ludens*) proved to be highly erratic (more so than any other compound tested on it). All flies initially exposed to DDT in concentrations ranging from 2 to 4 pounds were killed, but after an extended period a small number of flies that immigrated into the study site withstood exposure to DDT at all concentrations up to 8 pounds. In contrast, 2 pounds of tartar emetic was considerably more toxic than any of the DDT concentrations tested.[24]

After notice from the Orlando laboratory of the BEPQ, the Tennessee Valley Authority through its Health and Safety Department inaugurated another important DDT testing program in 1943.[25] This department was prepared for the study by its extensive experience in spraying for malaria control in the southeastern states. During the summer of 1943, the TVA conducted field tests to determine the practicability of DDT larvicidal dusts for the control of *Anopheles quadrimaculatus* (which, along with various *Aedes* mosquitoes, carried the malaria plasmodium). After several joint meetings with the technical staffs of the Orlando laboratory, the TVA expanded its research program to include laboratory and field studies on house spraying and to investigate the effectiveness of DDT as an anopheline larvicide (killed juveniles) and adulticide (killed adults) when distributed by airplane as a dust, a spray, or a thermal aerosol.

With the advice and assistance of scientists from government agencies and universities, the TVA tested the new insecticide extensively. The tests fell into several categories: acute toxicity, residual toxicity, repellency of DDT-treated surfaces, use of DDT as larvicide (distribution by boat and air), DDT as an adulticide, and toxicity with fish and fish food organisms. Several interesting findings emerged from this research. First, a comparison with pyrethrum (a pesticide created from the stigmas of chrysanthemum flowers) showed DDT to be nearly an order of magnitude less toxic to male and female mosquitoes than pyrethrum, which was the primary alternative in mosquito control. However, DDT showed very high residual toxicity (barns treated with 200 mg DDT/ square foot remained almost entirely free of mosquitoes for eleven weeks,

DDT AND ENVIRONMENTAL TOXICOLOGY

while occupied houses treated with 250 mg DDT/square foot remained free of *A. quadrimaculatus* for at least three months). Still more striking was the discovery of an even higher degree of residual toxicity to house flies. Finally, scientists found that DDT dusts and thermal aerosols gave no evidence of injury to fish or other aquatic organisms when applied by plane at rates of 0.1 pound per acre. A 5 percent solution of DDT in kerosene applied at approximately 0.25 pound DDT per acre, however, largely destroyed aquatic insects living in close contact with the water surface (notably Hemiptera and Coleoptera).[26]

From the results of this early research into the effects of DDT on mosquitoes, it was possible to conclude that the new insecticide was not the most toxic of available insecticides (pyrethrum appeared to be much more toxic to *Anopheles* mosquitoes according to early tests). Nevertheless, DDT showed great promise as a "magic bullet" in the battle against mosquitoes, malaria, and other diseases. The new insecticide destroyed 90 percent of all *Anopheles* mosquitoes in many of the tests, whether conducted indoors or outdoors. That figure was impressive to economic entomologists, but the difference between 90 percent and 100 percent would become extremely significant when DDT became available for widespread use.[27] Along with high toxicity, DDT showed a high level of persistence in most environments (another factor that impressed economic entomologists).

Beyond the toxicity of DDT to target organisms (e.g., *Anopheles quadrimaculatus*), scientists at the TVA considered potential side effects, such as the impact on nontarget insects and fish. At least as significant as the study of nontarget organisms was the range of environments examined. TVA scientists evaluated the effect of DDT in houses (occupied and unoccupied), fields and forests, and bodies of water.

Percy N. Annand (chief of the BEPQ) drew heavily on the above experiments in his address "How about DDT?" to the annual meeting of the National Audubon Society in October 1945, in New York City. Annand argued that insects posed a serious threat to forests in the United States and that he had witnessed the destruction of more than three billion board feet of lumber wrought by the spruce bark beetle in Colorado over the span of three years. After listing many of the insects investigated in the studies by the bureau and other agencies, Annand acknowledged that DDT produced unsatisfactory results against other

45

species, but on the whole, he downplayed potential risks associated with DDT. He suggested that it could be applied annually for many years before harmful quantities would build up in the soil, even if it did not decompose. As for the toxicity of DDT, Annand noted that the FDA would set the tolerance in fruits for DDT at 7 mg/kg, which was similar to the tolerance in fruit for lead and fluorine.[28]

Annand recognized the need for more research on the toxicity of DDT to farm animals and noted that tests were under way at several experimental research stations. The threat to honey bees concerned him, but compared to the effects to arsenicals, he thought DDT would be considerably less toxic to bees and that much of the early alarm was unwarranted. Likewise, Annand dismissed concerns about insect parasites and predators: "Except in operations far more extensive than any which are contemplated, it would appear that, even though all the parasites in an area were killed, it soon would be repopulated by infiltration from the outside."[29] Moreover, the chief of the Bureau of Entomology believed that the desire to maintain the balance of nature was unrealistic: "As a matter of fact, there are probably very few cases in which nature is balanced, and certainly it is grossly out of balance when there are extensive outbreaks of insect pests."[30]

After exploring the effects of DDT on wildlife, Annand concluded optimistically: "As our experimental work progresses, we are much impressed by the very small amounts of DDT that are effective against injurious insects. In the case of forest insects, particularly defoliators, excellent control has been obtained with ¼ to 1 pound of DDT per acre. The gypsy moth, cankerworms, and sawflies seem particularly easy to control."[31] Thus what distinguished DDT from other insecticides, such as pyrethrum and lead arsenate, was the remarkably low amounts needed to control target insects. In Annand's view, these quantities were too low to pose risks to other wildlife, whereas one had to apply much greater quantities of other non-synthetic insecticides to control insects effectively. By extension, Annand also believed that DDT could be applied at low rates and that it would accumulate much more slowly in the environment. Such a statement can be seen as comparative (earlier forms of insecticides accumulated in the environment rapidly). In his statements to the National Audubon Society, Annand clearly distilled the hopes and expectations of economic entomologists and public health officials for DDT.

DDT AND ENVIRONMENTAL TOXICOLOGY

As we have seen, economic entomologists celebrated DDT for its high toxicity against a wide range of insect pests. Few questioned the toxicity of DDT when employed in the fight against insects, yet scientists wondered if a chemical that was highly toxic to insects might also be to some degree toxic to plants, animals, and even humans. To explore this possibility, they conducted extensive pharmacologic studies on laboratory animals, including rats, rabbits, chicks, mice, guinea pigs, dogs, and monkeys. The results suggested the potential risks DDT posed to humans. Both the FDA and the PHS undertook ambitious research programs in this direction.

In 1944, Maurice I. Smith (chief pharmacologist at the PHS) and Edward F. Stohlman (associate pharmacologist) published the results of one of the earliest studies of the pharmacologic action of DDT on tissues and body fluids. Recall that Smith also conducted the toxicological analysis on Jamaica ginger in 1930. To test DDT, they devised a method to estimate the quantity of the chemical in its pure form. Noting that DDT given in aqueous suspension was irregularly and poorly absorbed in the stomach and intestines, the scientists administered DDT to rats and rabbits in olive oil, which produced superior results. For rats, the LD_{50} for DDT given intragastrically in 1 percent to 5 percent solution was 150 mg/kg (of body weight) and for rabbits 300 mg/kg, but death could be delayed for several days. DDT proved to be three times as toxic as phenol in rats and possibly twice as toxic to rabbits. Clearly, its acute toxicity was fairly modest, but scientists also considered the possibility that chronic toxicity might also be a serious problem.

Smith and Stohlman noted: "The effects of DDT in experimental animals are cumulative, and small single doses given repeatedly lead to chronic poisoning. In a group of 10 rats of about 80 grams weight, DDT fed at a level of 0.1 percent in a semisynthetic adequate diet containing 18 percent protein as casein was uniformly fatal in from 18 to 80 days."[32] Chronic toxicity was similar for rabbits and cats, although they were tested differently.[33] Two cats that received 50 mg/kg every day or every second or third day developed all the characteristic symptoms of poisoning and died in one case after twelve days with a cumulative dose of 500 mg/kg and in the other after fifteen days (cumulative dose of 300 mg/kg).

Having established levels for acute and chronic toxicity of DDT, Smith and Stohlman examined whether DDT could be absorbed

47

DDT AND ENVIRONMENTAL TOXICOLOGY

through the skin. They found little dermal absorption of DDT from impregnated clothing, but DDT applied to the skin in a dimethylphthalate solution was definitely toxic (dimethylphthalate was a nonirritant with a low toxicity to rabbits when given orally). Yet the scientists failed to rule out skin adsorption of DDT from impregnated cloths as approximately half of the animals exhibited systemic effects on the central nervous system.[34] Thus one of the earliest pharmacologic investigations of the toxic effects of DDT reached a striking conclusion: "The toxicity of DDT combined with its cumulative action and absorbability from the skin places a definite health hazard upon its use. Symptomatically the effects on the central nervous system are the most obvious, damage to the liver is less obvious and for this reason perhaps more serious."[35]

M. I. Smith also collaborated with R. D. Lillie (senior surgeon of the PHS) to determine the pathology of the poisoning experiments described above. This study clarified some of Smith's findings. Despite Smith's observations of pronounced neurologic symptoms, pathological examination revealed only slight histologic alterations in the central nervous system. As Smith suspected, the liver contained the most striking pathologic alterations. Hyaline degeneration was similar to that described in poisoning by azo-benzene (Lillie, Smith, and Stohlman had previously published their study of azo-benzene). Lillie and Smith also observed a variable amount of fatty degeneration of liver cells (often centrolobular) in cats, rats, and rabbits (all of the animals considered in this study).[36]

In 1943, a team of researchers from the NIH that included Neal (senior surgeon) and Oettingen (who had become principal industrial toxicologist) conducted experiments to determine the toxicity and other potential dangers of aerosols containing DDT. The research was divided between the Industrial Hygiene Research Laboratory and the Pathology Laboratory. They exposed animals for forty-five minutes to approximately 1 pound of DDT aerosol in a glass chamber. Guinea pigs, rats, and puppies showed no signs of discomfort or poisoning during or following exposure to initial concentrations of 54.4, 12.44, and 6.22 mg of DDT per liter of air over the period described. Mice, however, became "jumpy" shortly afterward and subsequently developed tremors, became hyperexcitable (as in strychnine poisoning), and developed clonic-tonic convulsions shortly before death. Eleven out of twenty mice exposed

48

to an initial concentration of 12.44 mg of DDT per liter of air died during the four days after exposure. Yet the remaining nine mice showed no toxic symptoms and gained weight during the following three weeks. When the researchers dropped the initial concentration of DDT slightly (from 6.2 to 6.1 mg/liter), none of the animals (including mice) exhibited toxic symptoms. Raising the concentration of sesame oil in the aerosol to 9.5 percent induced toxic effects in the mice, prompting the researchers to conclude: "It is, therefore, apparent that the toxicity of DDT for mice may be increased by increasing the percentage of sesame oil in the aerosol mixture."[37] The researchers attributed some of the effect of sesame oil to contamination of the mouse fur (ingested when the mouse licked its fur) and the toxicity of DDT was increased by the presence of fatty material.

The NIH team also tested the potential chronic toxicity of DDT aerosols, with experiments similar to those used to test for acute toxicity. Two puppies were exposed to a 1 percent DDT aerosol at a concentration of 12.2 mg/liter for forty-five minutes on two consecutive days one week and four successive days in the following week. Neither puppy exhibited any signs of poisoning, and both gained weight. As in the tests of acute toxicity, mice fared worse than puppies. Thirty mice were exposed to the same conditions as the puppies. Ten mice wrapped in gauze to protect their fur from contamination showed no symptoms of poisoning, but ten unprotected mice died with typical symptoms of DDT poisoning. Finally, ten mice with protective collars (to restrict licking of their fur) manifested typical symptoms that were slightly delayed and less severe, but most of the group died within five days following the exposure. A second series of tests refined the experimental design. In these tests, two monkeys and ten mice were exposed to a concentration of 0.183 mg/liter of DDT daily for a total of five weeks' duration. This concentration was less than 6 percent of that used for the initial tests of chronic toxicity. Neither the monkeys nor the mice exhibited toxic symptoms. Moreover, the monkeys gained weight (approximately 900 grams) and so did the mice (5 grams each on average).[38]

In effect, the tests for chronic toxicity amounted to acute toxicity tests stretched out over a longer duration as the researchers made no effort to identify etiology independent of acute symptoms. The PHS also tested humans for the toxicity of DDT (I will discuss these experiments, below, with the other tests conducted on human subjects). Researchers ignored all of these problems when they concluded: "Therefore, it may be concluded that in spite of its inherent toxicity the use of DDT in 1- to 5-percent solutions in 10 percent cyclohexanone with 89 to 95 percent Freon, as aerosol, should offer no serious health hazards if used under conditions required for its use as an insecticide."[39]

Neal and the NIH researchers also compared the toxicity of DDT inhaled to that ingested. Only one of three dogs in the inhalation group developed definite signs of poisoning. After closely monitoring this individual's symptoms, scientists found by autopsy that it had cirrhosis of the liver, prompting them to suggest that liver and kidney dysfunction may precede the onset of nervous symptoms. The three dogs that ingested DDT showed no distinct toxic manifestations, and autopsies of these dogs exhibited no gross pathological changes. After comparing the results of the two experiments, scientists concluded that dogs tolerated comparatively large doses of pure DDT (100 mg per kg) in capsules by mouth and by insufflation (forced inhalation) of the dry powder.[40]

This set of experiments addressed the crucial factor of body weight, but the techniques used to administer DDT may have affected the results. In the insufflation experiments, researchers blew DDT powder directly into the nostrils of dogs. In the ingestion tests, dogs were fed large quantities of DDT in capsule form. Neither set of experiments was likely to result in an exposure that could be replicated under noncontrolled circumstances. Yet the experiments did raise the possibility that sufflation of large doses of DDT caused definite signs of poisoning, preceded by injury to the liver and kidneys. Such findings broadened the range of possible toxic effects of DDT.

Determining the toxicity of DDT was not the sole purview of the NIH. When Congress cut funding for the FDA, explicitly prohibiting it from conducting research on arsenic and lead, it transferred responsibility for research on the toxicity of pesticides to NIH. Barred from research on lead and arsenic, the FDA pharmacologists shifted their emphasis to the toxicity of Elixir Sulfanilamide and glycols (see chapter 1). Next they analyzed the toxicity of mercury through an army contract. To avoid possible violation of the terms of the congressional appropriation, they undertook the study of the toxicity of DDT under the same contract.[41]

50

As early as 1943 the head of the Division of Pharmacy, Herbert O. Calvery, received a sample of DDT for toxicity testing. Calvery and a team of five scientists from the Division of Pharmacology tracked histo-pathological changes in 117 animals of nine species, including farm animals like chicks, dogs, cows, sheep, and a horse. As in Lillie and Smith's research, the most characteristic and most frequent lesion produced by the higher dosage levels of DDT was moderate liver damage. No one on the research team had seen the exact counterpart in any other experimental animals that they had ever studied. Because of the severe muscular tremors following large doses of DDT, researchers went to considerable lengths to obtain the brains and spinal cords from affected animals to search for evidence of physical changes to the central nervous system. Although they sectioned the brains of numerous animals and controls, they found no distinct differences between test and control animals. Finally, certain individuals were much more sensitive that others of the same species, but the lesions (i.e., manifestations of injury) were quite consistent throughout the different species.[42] In other words, differences between individuals of the same species were greater than differences between different species.

The FDA researchers also studied acute and subacute effects of DDT on small laboratory animals. Rats, mice, guinea pigs, rabbits, and chicks received DDT dissolved in corn oil via a stomach tube. One problem with the design of this experiment was the quantity of corn oil required to dissolve larger doses of DDT. Still, the experiments showed that relatively small doses of DDT intoxicated small animals and that the dosage-mortality curve was flat (that is, survivals and deaths occurred over a wide range of dosage). The researchers also concluded that rats and mice were more sensitive to DDT in single does than guinea pigs and rabbits and that DDT in solution was more readily toxic than DDT in suspension.[43]

For the tests of subacute toxicity, the FDA researchers fed DDT in different concentrations (ranging from 0.0 percent to 0.10 percent) to four different groups of rats for a period of one year. Several of the rats at the high dosage levels exhibited typical DDT symptoms after a few days and some died. The survivors at this level eventually developed symptoms and perished. The rats exposed to 0.025 percent DDT did not suffer increased mortality over the course of the fifty-two-week experiment. Autopsies of the dead animals most commonly revealed slight to moderate liver damage, occasional testicular atrophy, and some degeneration of the thyroid. Although researchers initially suspected that the rats were developing a tolerance to DDT, when they calculated food intake per kilogram body weight per day and plotted this against age, it became clear that the amount of DDT consumed per kilogram per day gradually declined with age.

The FDA researchers also investigated the impact of DDT on growth rate in rats in a paired experiment. Eight pairs of male rats and eight pairs of female rats were placed on experiment at the beginning of weaning. One member of each pair received 0.05 percent DDT dissolved in corn oil mixed with food, and the other received a control diet containing an equivalent amount of corn oil without DDT. After eleven weeks, the scientists terminated the experiment and concluded that feeding 0.05 percent DDT to rats did not appear to significantly slow the growth rate although the general trend of the average growth rate was downward.[44] The conclusions of this report transcended the results of any of the individual experiments and pointed to one of the aspects of DDT that toxicologists found most troubling, namely, the variability in individual susceptibility, which made it difficult to estimate the safely tolerated dose or exposure.[45]

Calvery's team of researchers also investigated the dermal or percutaneous absorption of the DDT in anticipation of its eventual use on human body lice. Acute toxicity tests exposed rabbits to 4 g/kg body weight of 5 percent DDT powder (talc). Neither dry nor wetted powder produced toxic symptoms. Having ruled out acute toxicity, the FDA researchers wondered if DDT in other solutions would prove equally "innocuous." To find out, they conducted a ninety-day subacute experiment.[46] They found DDT to be mildly irritating to intact and abraded skin, especially when applied by patch test or on the hands of operators who worked with it on a daily basis for nearly a year. In a carefully worded conclusion, the scientists recommended caution in the use of DDT: "The above data indicated that the unlimited use of DDT solutions on the skin is not free of danger; however, some solutions of DDT have been found safe for restricted use."[47]

Although most of the early DDT experiments addressed its acute toxicity, a team of researchers from the Pharmacology Section of the

DDT AND ENVIRONMENTAL TOXICOLOGY

Chemical Warfare Service explored the chronic toxicity of DDT in dogs. Daily doses of 100 mg/kg initiated "moderate, coarse tremor," which disappeared after DDT was withdrawn. Raising the dose from 150 to 250 mg/kg of DDT produced more severe neurological disturbances (intense tremors involving all muscle groups, aberrations in gait, exaggeration of the stretch reflexes), but symptoms diminished and disappeared a few days after the treatment ended. At higher doses (greater than 250 mg/kg), persistent neurological signs developed: "severe hypermetria [high stepping each step], rigid, hyper-extended and abnormally abducted legs, and aberrations in gait" (dogs could not walk in a straight line but proceeded in a zigzag fashion). Other symptoms included inability to eat, leading to severe weight loss and dehydration. Researchers were most concerned, however, that the tolerance of dogs to DDT declined markedly: a 40 mg/kg dose (which yielded few or no symptoms in a normal dog) produced severe symptoms in dogs on which they repeated the experiment. They concluded: "It was apparent from these observations that irreversible symptoms had been produced by the prolonged administration of DDT. Their clinical nature indicated that injury to the cerebellum might have played a major rôle in their production."[48] The researchers also found that DDT damaged the liver, sometimes only moderately but in some cases so severely that the animals died. DDT had no detectable effect, however, on renal function in dogs.

For the most part, these DDT investigations did not require methodological innovations. Most of the experiments involved the calculation of LD_{50}s and pathological examinations. One exception was the research of Edwin P. Laug, one of the pharmacologists in the FDA Division of Pharmacology. Laug developed a biological assay for the determination of DDT in animal tissues and excreta based on the toxic response of the housefly (*Musca domestica*). The great sensitivity of the housefly to DDT (and other chemicals) provided an excellent indicator. According to Laug, under proper conditions, the LD_{50} was on the order of 2.5 mg/kg bodyweight for houseflies.[49] He extracted the fat from the tissue of a DDT-poisoned animal, placed the extract in a flask, and introduced 100 flies, which picked up residues of the extract while walking around the flask. After a specified amount of time, Laug counted the number of living and dead flies and plotted a curve to determine the point at which half of the flies had died. Using this bioassay, which he referred to as his

53

DDT AND ENVIRONMENTAL TOXICOLOGY

"flyo-assay," Laug could determine quantities of DDT on the order of 2.5 ppm. In time, Laug's flyo-assay would be superseded by more refined chemical methods, but it was a useful technique during the early analysis of the toxicity of DDT and other novel chemicals.[50]

Most of the early tests of the acute or subacute toxicity of DDT on laboratory animals lasted for a few months at most. Even the tests for chronic toxicity continued less than six months. Edwin Laug and his colleague at the FDA, O. Garth Fitzhugh, changed that state of affairs when they conducted experiments on rats lasting for at least six months and up to two years. In one paired feeding experiment, rats received diets containing either 800 or 1,200 parts per million (ppm) for six months, after which all the animals showed characteristic symptoms of DDT poisoning. After six months, tissue analysis revealed measurable amounts of DDT in all tissues (save two kidneys). DDT seemed to be particularly concentrated in the perirenal fat (i.e., surrounding the kidney), at 50 to 100 times as great as in any of the other tissues. The most compelling finding of this study came from a two-year exposure to DDT, after which there was a definite correlation between tissue level, most clearly seen in perirenal fat and the level of DDT consumed. Yet at 800 ppm, the DDT concentration in perirenal fat was of roughly the same order of magnitude after six months' exposure as after two years of exposure. In contrast, DDT levels in the kidneys continued to rise during the two years, reaching levels four to five times as high as that recorded after six months.[51] DDT was the first insecticide subjected to long-term studies lasting up to two years despite the fact that heavy metal insecticides like lead arsenate were arguably more toxic. Studies of the toxicity of lead arsenate to dogs and rats preceded the DDT studies by a few years even though the insecticide had been in use for decades. Not surprisingly, FDA pharmacologists conducted this research.[52]

Because of its potential use in dairy barns, an early concern was whether or not DDT was eliminated in milk. One study addressed this concern through a series of experiments involving rats and goats. Researchers fed a mixture of 0.1 percent DDT in chicken mash to three young female rats, each of which was nursing a one-day-old litter. Typical DDT tremors appeared in the adult rats between days 6 and 13 and in the young between days 14 and 15. After day 18 all of the rats had died except one adult and one juvenile. In another experiment, nine adult

54

rats fed solely on goat's milk received daily oral dosages of 1 gram DDT per 8 to 9 pounds (3.63 to 4.08 kg) body weight. All the rats died after 2 to 29 days with symptoms of DDT poisoning. Still another experiment demonstrated that nursing rats developed symptoms of DDT poisoning shortly after their mothers began to feed only on milk from the goats regularly fed DDT. A kitten also died with typical DDT symptoms after consuming the milk of a goat fed DDT for 25 days, but an unweaned baby goat showed no signs of poisoning despite freely suckling from a goat under treatment for 27 days. Researchers also wondered if the milk of goats regularly sprayed with DDT would become contaminated. In light of these mixed findings, researchers urged caution, particularly regarding the milk of cows: "The data strongly suggest the need for more intensive research on the toxicity of milk from dairy cows ingesting DDT residues either from sprayed or dusted forage plants or from licking themselves after being sprayed or dusted with this insecticide."[53]

The precautionary principle would suggest that those who stand to profit from the sale of a good like a chemical insecticide should demonstrate its safety before releasing it to consumers. Lack of premarket testing (except for target organisms) left regulatory agencies like the FDA and PHS scrambling to determine the toxicity of DDT.[54] Laboratory tests raised numerous issues that called for further study. Although the acute toxicity of DDT was, as expected, quite low, chronic toxicity experiments suggested that in the long term, DDT could pose serious threats. Certainly the findings of the FDA pharmacologists in the Division of Pharmacology, drawn from extensive studies that addressed acute and chronic toxicity, effects on various tissues, feeding and inhalation exposures, as well as metabolic function, concentration in fat cells and milk, significantly undermined the contention that DDT was completely harmless to warm-blooded animals. There was enough ambiguity in their findings, however, to suggest that DDT used judiciously did not pose a great threat to mammals. Early wildlife studies were no more conclusive.

The path from the laboratory to the field was fairly direct. During spring 1945 scientists began to examine wild animals for the effects of DDT. Researchers fed field mice various concentrations of DDT from 0.40 percent to 0.01 percent and 0 percent (as a control). Neither the control animals nor the mice receiving low doses (0.01 percent to 0.10

percent DDT) exhibited toxic effects. At 0.20 percent, however, two mice died before the end of the experiment (thirty days). Doubling that dosage killed four mice within nine days (a fifth mouse died on the twenty-first day). White-footed mice appeared less susceptible to DDT than field mice in similar tests. Inhalation experiments were less conclusive than the ingestion tests. Ten field mice placed in an artificial habitat sprayed by hand with a DDT oil mixture showed no evidence of toxicity even though DDT-sprayed oats were introduced on the seventeenth day.[55]

Wild cottontail rabbits were also highly sensitive to DDT in the laboratory. Four rabbits fed crystalline DDT developed tremors. Two died (one on the fifteenth day and one on the twentieth day of the experiment). In another series of toxicity tests with cottontail rabbits, DDT was administered through a stomach tube in six dose levels from 500 to 2,500 mg/kg bodyweight. No symptoms appeared in rabbits exposed to levels below 1,500 mg/kg, but one rabbit at the 1,500 mg/kg level showed tremors in the second day (but recovered). At the other levels, two out of four rabbits at the 2,000 mg/kg level died on the third and thirteenth days, respectively, and two out of three rabbits at the 2,500 mg/kg level died on the seventh and twelfth days, respectively.[56]

Like mammals, birds reflected the effects of DDT in laboratory studies. Don R. Coburn and Ray Treichler, biologists at the U.S. Fish and Wildlife Service, fed five-week-old bobwhite quail a mash diet containing DDT at percentages ranging from 0.40 percent to 0.005 percent. All experimental quail fed mash containing 0.05 percent or more of DDT died. Even at .025 percent, half of the bobwhite quail perished, and there were even deaths at the lowest percentage used in these experiments (.005 percent). To determine the acute toxicity of DDT to bobwhite quails, the FWS researchers administered single doses of DDT either in crystalline form or in a vegetable oil solution. For the crystalline form, dosages ranged from 50 mg/kg body weight to 1,000 mg/kg body weight, while the dosages for the oil solution ranged from 40 mg/kg bodyweight to 1,000 mg/kg body weight. These tests revealed that 200 mg/kg bodyweight was required to cause significant mortality. Unlike researchers with a background in pharmacology, the FWS biologists did not determine LD_{50}s for quail, but they were able to establish rough estimates of acute toxicity. Finally, Coburn and Treichler noted that the symptoms of DDT poisoning found in other animals were the same in

DDT AND ENVIRONMENTAL TOXICOLOGY

birds: excessive nervousness, loss of appetite, tremors, muscular twitching, and persistent rigidity of the leg muscles.[57]

As a class, wild amphibians exhibited high levels of sensitivity to DDT. After collecting wood frog egg masses from bottomland ponds along the Patuxent River, Lucille Stickel, another biologist with the FWS, separated the tadpoles into separate aquarium jars at the rate of 100 per jar. Stickel left control jars untreated and treated others with oil only. A third group of jars received DDT and oil at a rate equivalent to 5 pounds per acre. All tadpoles treated with DDT died within three to five days, while the controls and tadpoles in jars treated with oil alone remained healthy.[58]

Wild-caught fish exhibited slight if any reaction to DDT in early experiments. In one early study, Eugene Surber (another FWS biologist) stocked 100 brook trout, 100 rainbow trout, and 100 bluegill sunfish in mid-August in four connected raceways. Researchers sprayed with an oil solution of DDT at the rate of 1 pound per acre and observed that the DDT remained on the surface of the water for at least four hours. None of the brook trout or the rainbow trout died or showed signs of contamination, but 4 to 12 percent of the bluegill sunfish died within five days.[59] One potential problem with the design of this study is that DDT in an oil solution may not have dissipated beyond the surface of the water, meaning that only the fish coming into regular contact with the surface would have been significantly exposed.

Surber conducted another experiment, which specifically addressed this problem by stocking each of twelve small, hard-water ponds with fingerlings (50 bluegills and 50 largemouth bass). Three ponds received no treatment to serve as controls. Then he applied DDT in three forms to the other ponds: in an oil solution, an emulsion, and a suspension. After one week, FWS researchers drained all of the ponds and counted the surviving fish. DDT in suspension killed very few fish while DDT in solution killed 50 to 60 percent of the bluegills but very few bass. DDT in emulsion killed all fish of both species. In yet another experiment, Surber tried to determine if fish would be killed when fed only on DDT-contaminated flies. With his colleagues, he stocked three ponds with 25 adult and 25 fingerling bluegill sunfish. Fish in all three ponds could gorge themselves on flies. In two of the ponds the flies had been sprayed with a 12 percent solution of DDT at the rate of 1 pound per acre, and in

57

the third pond the flies were untreated (as a control). None of the fish in any of the ponds died.[60] The impossibility of recording symptoms in fish, other than death, however, prevented these experiments from detecting subtler effects of chronic DDT poisoning.

Several conclusions emerge from the laboratory tests conducted on wild animals. With the exception of the tests on field mice and rabbits (both groups with lab analogies), they bear little resemblance to the tests conducted on lab animals. Experiments with birds, fish, and amphibians had to be created, de novo. On the whole, the size of the samples was significantly lower in laboratory studies of wildlife. While lab scientists noted considerable variation in individual susceptibility to the toxicity of DDT, wildlife biologists began to track considerable variation in the susceptibility of the many species of wildlife. These factors would all intensify as wildlife biologists moved from the laboratory to the field to study the effects of DDT.

In 1946, two scientists with the FWS, Clarence Cottam (assistant director) and Elmer Higgins (chief of the Division of Fishery Biology), captured the expectations for, and fears of, DDT in one of the first reviews of its effects on fish and wildlife (most of which had not been published at the time of the review): "From the beginning of its wartime use as an insecticide the potency of DDT has been the cause of both enthusiasm and grave concern. Some have come to consider it a cure-all for insect pests; others are alarmed because of its potential harm. The experienced control worker realizes that DDT, like every other effective insecticide or rodenticide, is really a two edged sword; the more potent the poison, the more damage it is capable of doing."[61] In discussing the potential hazard of DDT to wildlife, Cottam and Higgins emphasized the concept of specificity in toxicity: "Most organic and mineral poisons are specific to a degree; they do not strike the innumerable animal and plant species with equal effectiveness; if these poisons did, the advantage of control of undesirable species would be more than offset by the detriment to desirable and beneficial forms. *DDT is no exception to this rule. Certainly such an effective poison will destroy some beneficial insects, fishes, and wildlife.*"[62] But what risk did extensive use of DDT pose to wildlife?

To determine the extent to which DDT contaminated the environment, scientists sprayed a DDT solution at the rate of 2 pounds of DDT per acre from an airplane on a 117-acre tract of well-drained forest on the Patuxent River bottomland in the Patuxent Research Refuge in Maryland. By placing petri dishes (four inches in diameter) throughout the area prior to spraying and running them through chemical analysis afterward, researchers discovered that the amount of DDT reaching the ground through the forest canopy represented a small fraction of the original deposition (0.008 pound per acre under tree canopy and ground cover, 0.5 pound per acre on open forest floor under tree canopy, and 0.6 pound per acre under tree cover along a riverbank). Insect control was not an objective of the study, but researchers noted that many insects died within a few days of application of the spray, particularly adult mosquitoes. The effects were only temporary; most species returned to normal numbers in two to three weeks.[63]

Scientists at Patuxent Research Refuge (a National Wildlife Refuge) tracked the effects of DDT on mammals, birds, amphibians, and fish. Using 50 live traps, researchers counted the number of two species of insect-eating mammals (the short-tailed shrew and the deer mouse) on a 10-acre sprayed area and on a similar 10-acre control area just over a mile from the sprayed tract. The populations of both mammals declined in both sites, so the findings were without statistical significance: "The differences on the two areas are not of statistical significance, and the consistent reductions may be due to seasonal changes in behavior."[64] For the bird studies, FWS scientists conducted an intensive search for nests before they sprayed a 31-acre area within the 117-acre tract of bottomland forest. They also made censuses of two additional areas: a 22-acre area adjacent to the sprayed area and a 32-acre area slightly more than a mile away. Only one species, the American redstart (*Setophaga ruticilla*) declined significantly, possibly because as a tree-top feeder it suffered greater exposure.[65]

Another experiment attempted to determine if spraying DDT at the rate of 5 pounds/acre to birds' nests would disrupt the hatching of eggs, disturb the development of young, or cause the abandonment of eggs. Using a hand atomizer, scientists sprayed DDT on a one-square-foot area surrounding and including a nest. After attempting to compare pairs of nests of the same species, researchers concluded: "The treatment with DDT showed no detrimental effect on the hatching of eggs or on the development of the young; it caused no abandonment of nests even when they were located in such confined quarters as bird boxes."[66]

In the same sites as those used for the bird studies, another researcher studied the effects of DDT on frogs and toads in the wooded bottomland. In one experiment, various species of frog and toad tadpoles were stocked in open-topped cages and inspected daily for nine days after spraying. None of the animals were visibly affected by the spraying. In another experiment, a scientist treated two artificial ponds inhabited by adults and tadpoles of several species of frogs and toads with xylene and fuel oil only, two with DDT at 1 pound/acre in an oil solution, and two at 5 pounds/acre in an oil solution. Several untreated ponds served as controls. One pond of each pair was deeper than the other. Researchers sampled all of the ponds with dip nets twice prior to spraying and several times after. None of the amphibians in the deeper pond treated with 1 pound/acre DDT died, but several frogs and large-frog tadpoles as well as a young water snake died in the shallow pond (five inches deep at the center). Several additional frogs died in both of the ponds treated with 5 pounds/acre DDT. Nevertheless, some amphibians survived in all ponds. Researchers attributed all deaths to DDT.

As in the laboratory experiments, fish were highly sensitive to DDT. Nearly a mile of the Patuxent River (a muddy stream with a flow of about 130 cubic feet/second during the summer) passed through the 117-acre tract that was sprayed with DDT. Only 9.5 hours after the initial spraying, researchers seined 95 dead fish out of the stream near the lower end of the sprayed section. Dead fish continued to drift into the stop net four days after treatment, but the greatest number were lost during the first two days. In what was in effect a controlled experiment, scientists stocked eight shallow, soft-water ponds (20-by-50 foot) with several fish species. After spraying the three groups of ponds with three concentrations of DDT (0.1, 0.5, and 1 pound/acre) and leaving one pond unsprayed as a control, researchers found that mortality was considerable in all ponds but most severe in the pond treated with 1 pound/acre.[67]

The majority of the wildlife studies discussed so far had little or no connection with laboratory studies, except for the several studies of wildlife in the laboratory, which had made a concerted effort to mirror laboratory experiments. Overall, the techniques, language, and approach to experimental design shared little with the lab. The reasons for the marked differences between the lab and the field will emerge in a closer study of the major field studies of the effects of DDT on wildlife.

As with the laboratory studies, several agencies independently evaluated the toxicity of DDT to wildlife. One of the most extensive analyses of the potential impact on nontarget organisms was undertaken by the PHS at the Carter Memorial Laboratory in Savannah, Georgia. Directed by Clarence M. Tarzwell, a retired senior assistant sanitarian at PHS, this study sought to explicate the effects of DDT mosquito larviciding on wildlife. More specifically, the purpose of the studies "was to determine at what dosages and in what manner or physical state DDT could be routinely used as an anopheline larvicide without being significantly harmful to other organisms of economic or recreational value."[68] In late 1944, the first year of the study, researchers conducted experiments on the effects of routine hand application of DDT dusts, emulsions, and solutions, varying the method of application, types of larvicides, and dosages of DDT. Early in the study, they discovered that tight emulsions and solutions of DDT applied at a rate of 0.4 pound/acre were deleterious to the fish population in shallow waters. For this reason, they shifted their emphasis to DDT dusts or solutions at lower concentrations (0.1, 0.05, or 0.025 pound/acre). At these levels, they observed no fish mortality in individual treatments, but repeated routine treatments caused fish to begin dying in the interval between three and ten treatments, and eleven to eighteen treatments at this rate significantly reduced the population.[69]

By 1945 the PHS had significantly expanded its assessment of the toxicity of DDT. Researchers studied the effects of the routine treatment at 0.1 pound DDT per acre, applied by airplane to extensive areas of the Savannah River National Wildlife Refuge in Georgia. The initial emphasis on fish and fish food sources (surface, bottom, and plankton organisms), expanded with the assistance of the FWS to include studies of the effects of routine treatments on amphibians, reptiles, birds, mammals, and terrestrial insects. Spraying continued during a third season (1946) as researchers considered the cumulative effects of two years of routine treatments on the fish and wildlife populations.[70]

For this research, scientists examined ponds in three areas of the Savannah River Refuge and fourteen natural ponds at the Plant Introduction Laboratory of the Bureau of Plant Industry. They sprayed DDT at weekly intervals (routine treatments) in various concentrations ranging from 2 pounds to 0.025 pound per acre, but most experiments fell in the range of 0.1 to 0.025 pound per acre. Scientists used two methods to detect kills or changes in the population of surface organisms. The first method was gross observations taken 24 to 48 hours after treatment to detect any kill of the larger insect forms (such as Gyrinidae, Dytiscidae, Hydrophilidae and Corixidae). The second method was to take quantitative surface samples before and after treatment in order to determine any changes in the population of surface organisms. Researchers soon discovered a significant problem with their methodology. After collecting twenty-five random samples at the beginning of the study and before and after treatment, they realized that there were no large homogeneous areas suitable for such sampling in the ponds, and that the numbers of organisms found in the different samples varied considerably. Further, they noted that in most instances the variation was so great that it would have been impossible to detect even large differences due to treatment. Consequently, they abandoned random samples in favor of paired samples.

In the absence of treatment, there were no significant differences in the samples taken, which clearly affirmed that the paired sampling technique was valid. Scientists made a consistent effort to reduce variation by rigidly controlling the sampling technique so that differences due to the treatment might be detected. To determine the significance of the differences in the samples, researchers employed the student's t-test. They used P values to denote levels of significance (with a value of 0.05 or less considered significant). They clarified the effects of individual treatments with these methods and demonstrated residual or cumulative effects due to routine treatment by comparing graphically the populations in the treated and check (control) ponds throughout the period of treatment.[71] Of all the wildlife studies of the effects of DDT, the PHS's larviciding experiments most faithfully represented the most sophisticated ecological methods at the time.[72]

Careful methodology produced clear results. Individual treatments in shallow ponds with a sand bottom at rates of 1 to 2 pounds of DDT per acre killed numerous aquatic animals, including invertebrates like dragonflies and mayflies, as well as fish. Even after they reduced the spray concentrations, PHS officials recorded high mortality rates for

DDT AND ENVIRONMENTAL TOXICOLOGY

nontarget aquatic insects of many orders. Although kills were more pronounced at higher dosages, there were kills at all dosages (even slight
kills at 0.025 pound per acre after the first few treatments).

Meticulous observations clarified the effect of DDT on nontarget
aquatic insects. By comparing the effects of DDT spray to the effects of
fuel oil alone, the PHS scientists justified the high mortality rates: "The
over-all results suggest that 0.05 pound and 0.025 pound of DDT per
acre in fuel oil kills only a fraction as many surface forms as do applications at 0.1 pound per acre, and that fuel oil in itself kills numerous
forms."[73] The PHS officials suggested that treatments at low concentrations might prove less harmful to nontarget insects than traditional
methods of mosquito control: "It may be that 0.025 to 0.05 pound of
DDT applied in 1 gallon of fuel oil per acre will kill considerably less
insect life than the regular routine oiling at 15 to 40 gallons per acre
which has been used for mosquito control in the past."[74]

In addition to the gross observations of mortality, scientists took
quantitative surface samples in many of the treated and check ponds.
Although a few of the samples showed significant changes in the insect
populations in particular ponds, none of these changes were consistent
when ponds were compared. The greatest changes in the number of insects occurred in the untreated check pond, but researchers attributed
this change to sampling error. Nevertheless, scientists were able to map
certain population trends using the data from quantitative sampling.
Chironomids (midges) suffered the highest rates of mortality, particularly in a pond treated with a solution of fuel oil and 0.05 pound DDT
per acre. Yet the total population of surface forms increased in treated
areas. Specifically, nematodes, oligochaetes, and copepods increased in
the treated ponds at the wildlife refuge, suggesting that DDT reduced
their predators.[75]

Researchers hesitated to draw larger conclusions, for example, what
these changes might mean for other organisms or the ecosystem as a
whole. They did suggest, however, that some of the species that suffered
serious declines might represent sources of food for fish and that the
forms whose population exploded would not serve as a substitute food
source. Nevertheless, they qualified this pessimistic suggestion: "Reductions noted to date, however, have not been sufficient to affect the
breeding stock, and since treatment is in localized areas, it is probably

DDT AND ENVIRONMENTAL TOXICOLOGY

not sufficient to seriously limit the fish population by restriction of the food supply."[76] Other PHS biologists would take up the question of the effects of DDT on higher forms, but the potential impact of DDT on the food sources for fish remained central to their research.

Researchers also examined the effects of larviciding with DDT on birds and mammals at the Savannah River National Wildlife Refuge, beginning shortly after the study of aquatic invertebrates and fish and to a considerable degree sharing the same sophisticated methodology as that study. The study of birds and mammal effects concentrated on about 815 acres of the refuge, which were larvicided during the summer of 1946 with DDT (in a 20 percent solution in highly methylated naphthalene) at the rate of 0.1 pound of DDT per acre.

Before they sprayed, researchers mapped the ten islands in the study and set out numbered stakes in rows 100 feet apart with 100 feet between the stakes in each row to form a grid for recording the census of singing male birds and the live trapping of mammals. In an effort to build upon earlier research, which emphasized the analysis of the effects of single, high dose treatments of DDT, the PHS biologists modified their approach for light treatments on a regular basis over the extended period required for effective larviciding. Then they counted singing males, on a weekly basis, beginning on March 26, 1946, and continuing until August 8, when the breeding season had concluded for most species. Although the number of singing males increased on both sprayed and unsprayed islands, the arrival of new migrants contributed to this change. In both sprayed and unsprayed areas, the number of singing males rose from a low at the beginning of the nesting season, fluctuated slightly during the season, and fell as the end of the season approached. PHS biologists concluded: "The absence of a sudden drop or a gradual decline in the population of the sprayed area indicates that the DDT spraying did not affect the population to any appreciable extent."[77]

Mammals seemed to be similarly unaffected by routine larviciding with DDT. The most common mammal was the cotton rat (*Sigmodon hispidus*), a large herbivorous rodent, which was unlikely to feed on insects. Given that the rates of recapture were similar (21.1 percent on the unsprayed areas and 25.4 percent on the sprayed areas), researchers concluded that the activity of the rats and their rate of mortality were about the same on both areas, suggesting that DDT had no apparent effect on

64

DDT AND ENVIRONMENTAL TOXICOLOGY

the rodent population. Moreover, they trapped only half as many cotton rats on the unsprayed area as on the sprayed area, giving rise to a provocative conclusion: "Thus, DDT had no apparent effect on the reproductive potential of the rats on the sprayed area. Judging by trap catches, the potential was slightly greater in rats of the sprayed area."[78] Researchers also studied rabbits, raccoons, and cotton rats through sight observations on daily drives through the sprayed and unsprayed areas. From 174 observations of immature rabbits on the unsprayed areas and 244 observations on the larger sprayed areas, the biologists concluded that there was no significant difference after allowing for the greater size of the sprayed area. This information did not provide rigorous support for their conclusion, however: "DDT, then was not interfering with the reproductive capacity of the rabbits in the sprayed area as indicated by these counts."[79]

Despite the methodological rigor of the PHS studies of the effects of DDT larviciding on wildlife, the studies could not effectively evaluate nonacute or chronic effects of low concentrations of DDT on wildlife. As in so many other early experiments, these studies focused on a single endpoint—the death of the organism—and anything short of death failed to register in the analyses. In the years following World War II, toxicologists were just beginning to develop methods of analysis that would enable them to evaluate the effects of chronic exposures. Such effects included carcinogenicity and reproductive effects as well as subtle neurotoxic effects. In the case of DDT, scientists were unable to discern a pattern in the appearance of these effects, except that at high doses most animals experienced neurological symptoms prior to death. Fitzhugh did report tumorigenic activity in mice and others noted that DDT could be passed from mother to offspring via milk, but neither of these isolated reports created a clear picture of the toxicity of DDT to all organisms. Studies of the toxicity of DDT to humans were no more definitive.

Some of the earliest studies of the toxicity of DDT were conducted on human subjects. Scientists wished to rule out potential hazards of occupational and casual exposure to DDT before the chemical reached the general public. The NIH and the PHS took particular interest in the human health effects of DDT. Neal, Oettingen, and others conducted inhalation experiments on two adults. The subjects underwent extensive

65

tests before and after exposure. Tests included: pulse, blood pressure, respiratory rate, size of pupil and appearance of eyegrounds and conjunctivae and presence of nystagmus, inspection of the throat, steadiness tests (finger-nose test and steadiness of extended fingers), and biceps and Achilles reflexes, urine tests (to determine albumin, sugar, urobilin, urobilinogen, cellular constituents, pH, casts, and specific gravity), blood analysis (red blood cells, hemoglobin, white blood cells, and differentials), and psychophysiological tests (including mental alertness).

From this extensive battery of tests, Neal and others found no subjective systemic manifestations of DDT intoxication. In addition, results of the physical examination were negative, and the scientists recorded no physiological aberrations other than a drop in hemoglobin and red blood cell count. The NIH scientists summarized their results as follows: "The experiment shows, further, that inhalation of 3.1 to 4.0 mg of DDT in the form of DDT aerosol for 1 hour daily, on 6 consecutive days causes no subjective or objective symptoms in human subjects."[80] In a second series of experiments, researchers raised the exposure level to three times the original experiment. The same two subjects received the higher level beginning four weeks after the original experiment. Again, scientists completed an extensive medical examination before and after exposure. As with the first experiment, none of the tests revealed clinical signs of toxicity despite DDT deposits on the vibrissae (hairs) in the nose of one subject and coating the hands and forearms throughout the entire experiment. Thus the scientists concluded that an exposure to a total of 124.8 grams of a DDT aerosol "produces no toxic effects in human subjects and should offer no serious health hazards if used under conditions required for its use as an insecticide."[81]

Several important studies of the effects of DDT on human subjects were conducted in Great Britain. One researcher completed numerous experiments with volunteers wearing undergarments impregnated with 1 percent DDT (dry-weight basis). This research included also a small group of technicians engaged in laboratory work and bulk impregnation, which brought them into contact with DDT. None of the fifty-eight men manifested symptoms suggestive of toxic absorption, although a few had slight, transient attacks of dermatitis, which may or may not have been caused by DDT. From these results, the researcher concluded that soldiers under battle conditions could safely wear garments impregnated with DDT as a deterrent to lice. He also cited, without references, comparable studies of men engaged in spraying DDT conducted by Americans.[82]

Another British researcher placed two male subjects in an octagonal chamber (six feet across, six feet high) for two forty-eight-hour periods with an interval of forty-eight hours in between. The walls of the chamber were painted with a distemper (film). For the first period of exposure, the distemper did not contain DDT, but in the second exposure the distemper contained 2 percent DDT. The subjects wore shorts only and were forced to sit so that a large portion of their skin came into continued contact with the oily film on the walls. Medical examinations (conducted every twenty-four hours) included clinical neurological examination, electro-encephalograms, hematological examination, and urine analysis, as well as notes of subjective phenomena.[83]

The extensive examinations revealed no measurable changes during the control period, but there were significant changes during the period of exposure to DDT. Subjectively, both subjects felt eye pain, tiredness, heaviness, and aching limbs and became extremely irritable and disinterested in work of any kind. They felt unable to tackle the simplest mental task (although one subject was able to complete mathematical problems with normal precision). Both subjects suffered intense joint pain, and one had to spend a day in bed. In addition to these pronounced subjective findings, there were slight neurological effects: reflexes diminished, auditory acuity altered, one subject had peripheral anesthesia, and the other had a fine tremor.[84] This kind of intense exposure stimulated neurological effects that were not recorded in other studies of human subjects.

Another British study of the effects of DDT in humans was conducted at the Royal Naval School of Tropical Hygiene. Researchers followed the health of fifteen men attached to the school, who were for many months heavily and continuously contaminated with a 5 percent solution of DDT in kerosene. A variety of clinical and special studies (including renal and liver function, blood investigation, general demeanor, and labor output) did not show any ill effects associated with DDT. In another experiment, researchers exposed six human volunteers for 27.5 hours over five days in an experimental room to a continuous-phase aerosol of DDT. And in yet another experimental room they treated a

DDT AND ENVIRONMENTAL TOXICOLOGY

group of five subjects in the same way at night, and/or continuously for three months. None of the volunteers exhibited ill effects. Royal Naval scientists concluded, "DDT when used as an insecticide, with reasonable intelligence and the precautions normal to the use of modern insecticides, is harmless to man and animals."[85]

In the U.S., besides Neal's research, another scientist for the PHS conducted much of the analysis of the toxicity of DDT to humans. Based in Savannah, Georgia, Wayland J. Hayes, Jr., was chief of the Toxicology Section of the Technology Branch of the Communicable Disease Center of the PHS. In 1949, Hayes and two colleagues analyzed human fat for the presence of DDT. Previous studies had reported that both humans and animals excreted degraded DDT (metabolites) in urine.[86] No systematic study had been published save for the research of Laug and colleagues, who used a technique that made it impossible to determine the quantities of DDT (or metabolites) present in the original fat samples.[87] Hayes and his colleagues determined that the majority of samples contained a large proportion of degraded DDT (DDE, a metabolite of DDT), but they could not determine whether DDE was present because DDT residues degraded on plant products prior to ingestion, during digestion, or after deposition in human adipose (fat) tissue. Nevertheless, they called for a reconsideration of the possible health hazards associated with the widespread use of DDT.[88] Years later, when it became clear that DDT was wreaking havoc to the reproductive systems of birds of prey, DDE emerged as the culprit.

In late 1953, Hayes initiated a comprehensive analysis of the toxicity of DDT to humans. With prison volunteers as subjects, Hayes designed experiments to study possible clinical effects at different dosages, the relation between oral dosage and storage of DDT and metabolites in adipose tissue, and the relation between oral dosage and urinary excretion.[89] For periods of up to eighteen months, each of the fifty-one volunteers ("with full knowledge of the plan of the study and with complete freedom to withdraw at any time") consumed 0, 3.5, or 35 mg of DDT every day. Hayes set the dosages according to O. Garth Fitzhugh's estimate of the daily ambient quantity of DDT consumed by an average adult: 1.75 mg. Hayes set the volunteers' dosages at roughly 20 times and 200 times this amount. At the end of the study, Hayes and his colleagues commented: "During the entire study, no volunteer complained of any symptom or showed, by the tests used, any sign of illness that did not have an easily recognized cause clearly unrelated to exposure to DDT."[90] Hayes's research on the storage of DDT in humans showed that after one year humans reached a threshold, after which they accumulated no more DDT. He also found that human subjects excreted 20 percent of the DDT administered as DDA, another metabolite of DDT, like DDE, in their urine. Hayes and his colleagues concluded: "The results indicate that a large safety factor is associated with DDT as it now occurs in the general diet."[91] As in the case of so many of the early evaluations of the toxicity of DDT, scientists defined safety as the absence of acute toxicity or clinical effects. Nor did Hayes consider differences between various kinds of exposure. His ingestion study had no bearing on inhalation toxicity or dermal exposure. By 1956, when Hayes's study was published in the *Journal of the American Medical Society,* other scientists, doctors, and wildlife biologists alike were strongly criticizing DDT.[92] Despite such currents, Hayes continued to defend its use and safety.

Historian Edmund Russell has noted: "There were bombs before the atomic bomb, but the atomic bomb placed the attack against human enemies on a new plane. There were drugs before penicillin, but penicillin placed the attack against bacterial diseases on a new plane. There were insecticides before DDT, but DDT placed the attack against insect pests on a new plane."[93] Having examined in considerable detail many of the studies of the toxicity of DDT to target insects, laboratory animals, wildlife (including nontarget insects), and humans, we can now add that DDT not only transformed the attack on insects but also significantly influenced how scientists evaluated the toxicity of new chemicals. During and immediately after World War II, scientists scrutinized DDT intensely. It is safe to say that no chemical before had received such extensive study from such a wide range of scientists: economic entomologists, laboratory scientists, wildlife biologists, public health officials, and doctors.

Several trends link these disparate investigations. Most of the research focused exclusively on acute effects of DDT. Even the few studies that attempted to address chronic effects generally overexposed the test subjects to DDT (the one major exception being the two-year studies on mice, which also used fairly high doses of DDT). Because of the short duration of most experiments, few toxic effects developed—except at

DDT AND ENVIRONMENTAL TOXICOLOGY

high levels of exposure. Researchers concluded, therefore, that DDT was not harmful. Here the major exception seems to be laboratory studies in which numerous clinical and subclinical effects were noticed but dismissed or minimized in various ways.

Most researchers emphasized the direct effects of DDT and dismissed indirect effects, such as the elevation of populations of aphids and red mites, both of which survived exposure to DDT and thrived after their insect predators succumbed. Economic entomologists were, however, aware of this potential problem. The greatest variation in the individual experiments was in the concentrations of DDT used. Concentrations ranged from below 1 percent to greater than 10 percent, but the actual amount applied also varied from less than 0.1 pound per acre to 5 pounds per acre and more.

Turning to theoretical models for the analysis of DDT, we find that the laboratory studies drew heavily on pharmacology and its rigorous testing procedures (the laboratory studies were among the most consistent of all the early studies of DDT). The studies on the effects on target insects had a well-established model in economic entomology. Many of the wildlife studies did not reflect advanced ecological theory (one prominent exception was the joint research conducted by the PHS and the FWS on the effects of DDT larviciding). Finally, the human studies drew heavily on occupational exposure models and to a lesser extent on laboratory animal studies.

All DDT studies contributed in some degree to the study of toxicology. The laboratory tests refined the use of the LD_{50} as the benchmark standard of acute toxicity. In addition, the FDA pharmacologists strove to develop useful gauges of chronic toxicity, such as Laug's bioassay using houseflies as well as long-term studies that lasted up to two years (DDT was the first insecticide subject to such extended analysis). It is not entirely clear why biologists became involved in the analysis of the toxicity of DDT to wildlife. James Whorton demonstrated that wildlife did not concern the scientists who studied the pesticides that preceded DDT, such as lead arsenate.[94] But unlike lead arsenate, which was sprayed from the ground, DDT would be sprayed in aerosols from airplanes, thereby greatly expanding potential exposures to wildlife, including beneficial insects and vertebrates, among them humans. Nevertheless, DDT broadened the scope of toxicology by instigating wildlife

70

studies, which became part of toxicological evaluation. Studies of the effects of DDT on humans suggested little to no effects at low dosages or exposures, indicating an appropriate safety factor, but the accumulation of DDT metabolites like DDE necessitated further examination. DDT metabolites would later appear at toxic levels in the eggshells of birds of prey, one signal of extensive environmental contamination.

Toxicologists (qua chemists, pharmacologists, wildlife biologists, and physicians) rose to meet the new challenges posed by DDT. So novel was this technology that scientists struggled to find ways to identify and evaluate the risks it posed. The war effort coordinated and consolidated the work of many scientists in the study of the toxicity of DDT, just as the Elixir Sulfanilamide tragedy instigated concentrated toxicological analysis. But the study of the effects of DDT was not the only subject of wartime research; the army's Office of Scientific Research and Development sponsored many other scientific analyses of chemicals.

CHAPTER 3

The University of Chicago
Toxicity Laboratory

The discovery of DDT as an effective pesticide at the beginning of World War II resulted in extensive research as to its toxicity. DDT received more toxicological scrutiny—from entomologists, toxicologists, and wildlife biologists—during the first years of its release than any pesticide that preceded it. Yet the war and its aftermath produced many other new chemicals that called for toxicological screening. To assess the toxicity of these chemicals—and also their potential for wartime use—the army's Office of Scientific Research and Development (OSRD) awarded the University of Chicago a contract to evaluate these new substances. The University of Chicago is recognized for its great contributions to the war effort through the work of its renowned physicists, but university researchers also made significant contributions through the Toxicity Laboratory.[1]

The formation of the Toxicity Laboratory and the wide range of its research represents a central episode in the development of toxicology for at least two reasons. First, the Toxicity Laboratory was one of the first institutions devoted entirely to toxicological research, which was related to pharmacology but becoming increasingly distinct from it. Many prominent toxicologists began their studies or initiated their research at the Toxicity Laboratory. Second, although some of the broad research topics pursued therein seem distant from environmental risk and pesticides, there are important links in such areas as the joint toxic-

72

UNIVERSITY OF CHICAGO TOXICITY LABORATORY

ity of and resistance to antimalarial drugs, both factors that would be central in the analysis of insecticides.

As chair of the Department of Pharmacology at the University of Chicago, E. M. K. Geiling directed several simultaneous programs on behalf of the war through the Toxicity Laboratory. Geiling and his growing group of collaborators and students screened antimalarial drugs, evaluated the cancer-inhibiting effects of derivatives of the mustard gases, studied the fate of certain drugs through the use of radioisotopes, and explored the toxicity and pharmacology of organophosphate chemicals. Geiling had achieved considerable success as a faculty member at Johns Hopkins University and as professor and later chair of the Department of Pharmacology at the University of Chicago, where he oversaw critical studies. Federal authorities at the FDA were familiar with Geiling's research on Elixir Sulfanilamide (see chapter 1). The centrality of the University of Chicago in scientific efforts to support World War II made the incorporation of Geiling and the Department of Pharmacology a logical extension of ongoing research there.[2]

In 1936, the School of Medicine at the University of Chicago established a separate Department of Pharmacology and appointed Geiling as its first chairman. Geiling organized the department as an academic unit balancing teaching and graduate research. His first graduate student, Frances Oldham Kelsey, received her doctorate in 1938. Kelsey believed her admission to the graduate program might have been an oversight on Geiling's part, since it was not yet common practice to admit women, and Geiling addressed her acceptance letter to "Mr. Oldham" (Kelsey's maiden name). After Kelsey accepted the spot in the doctoral program, Geiling refused to admit whether or not he had been confused. Kelsey thrived as doctoral student then colleague before she joined the FDA.[3]

Well in advance of the formal entrance of the United States into World War II, a defense contract established the Toxicity Laboratory at the University of Chicago. The Orlando Laboratory for the study of medical entomology originated along the same lines. With the intensification of World War II, the National Defense Research Committee (NDRC) contracted the University of Chicago to establish a facility capable of evaluating the toxicity of chemical agents for the Chemical Warfare Service (CWS). In doing so, military officials hoped to avoid the crippling

73

UNIVERSITY OF CHICAGO TOXICITY LABORATORY

E. M. K. Geiling, preparing marine toad for bufagin extraction. Courtesy of the University of Chicago Photographic Archives apf1–06326, Special Collections Research Center, University of Chicago Library.

injuries inflicted on American troops by chemical warfare during World War I. One of the main reasons the NDRC selected Chicago was that the university had, in an old powerhouse, an unused smokestack, which could be used to ventilate the laboratory. In addition, Chicago had emerged as a center for research on the development of the atomic bomb. Finally, Geiling, recognized for his research on the toxicity of diethylene glycol (see chapter 1), was an ideal unifying force for the project.

With approval of the NDRC contract, Geiling became the principal investigator of the newly established Toxicity Laboratory on April 1, 1941, with Franklin D. McLean serving as the first director. During the war, Geiling and McLean developed the Tox Lab from a staff of six in a single building to a research cooperative numbering more than sixty investigators divided among seven large buildings. The physical size of the laboratory was a minor concern compared to assembling a skilled

74

UNIVERSITY OF CHICAGO TOXICITY LABORATORY

staff, defining research problems, and gaining the experience needed to interpret experimental results in terms of the needs of the armed services.[4] Lab scientists later recalled that at the outset they struggled to frame research questions in such a way that the results would contribute to the war effort. As a group, scientists were completely unfamiliar with tactical and strategic problems of the military. Nor were Chicago pharmacologists informed of the ongoing efforts of a small number of scientists in the laboratories of the CWS in Edgewood, Maryland, which left them isolated. Staffing was also a major challenge for the Toxicity Laboratory. As in the Manhattan Project, pharmacologists and other scientists were virtually conscripted from prewar jobs and graduate schools. On June 28, 1941, an executive order established the OSRD under Dr. Vannevar Bush. The OSRD comprised the existing NDRC and a new Committee on Medical Research (CMR).[5] Each of these groups negotiated contracts with many research universities and medical schools.[6]

In 1943, McLean resigned from the Tox Lab to accept a commission in the CWS and Dr. Keith Cannan from New York University became the new director.[7] Even as Geiling and other members of the Tox Lab confronted questions about research, space, and staffing, chemicals needing evaluation arrived by the dozen; eventually more than a thousand chemicals would arrive at a time. In order to analyze such a vast quantity of new chemicals, Tox Lab scientists were drawn from many scientific fields, among them biology, medicine, and physics. Numerous specialists, including pharmacologists, physiologists, biochemists, pathologists, chemists, physicists, mathematicians, ophthalmologists, and dermatologists joined the laboratory.[8] The new chemicals challenged Tox Lab researchers far beyond such initial questions as the chemical's toxicity when inhaled. Other questions included: "Did [a given chemical] cloud the eye or blister the skin as did mustard gas? Did it make a man cry, or sneeze, or his skin itch? And then, if it did any of these things, why? And what could be done about it? How good were the gas masks, the antigas ointments, and protective clothing?"[9]

The group evaluated the toxicity of several thousand potential chemical warfare agents, including nitrogen mustards, antimalarial drugs, radioisotope markers, and organophosphate poisons. Along with extensive laboratory space for researchers, the Toxicity Laboratory contained facilities for lab animals. The Tox Lab experiments used large numbers of monkeys,

75

dogs, rabbits, rats, and mice, not to mention cockroaches and silkworms, in toxicity studies. Animals were kept in the "zoology pens," which included dog runs. In contrast to most animal testing labs today, the pens were outside. The few trees provided the animals a minimum of shade but otherwise little protection, but the presence of dog runs suggested some concern for the well-being of the animals.[10]

To the leadership of the Department of Pharmacology, and in effect the Toxicity Laboratory as well, Geiling brought a dynamic style that inspired personal and professional loyalty in his staff. Never having married (and only grudgingly tolerant of marriage in his staff), Geiling treated the faculty, staff, and students of the Department of Pharmacology and Toxicity Laboratory like family, even personally selecting Christmas gifts (generally books that reflected an individual's interests). The responsibility of faculty promotions fell to Geiling, and each year when he submitted his recommendations to the dean, he dramatically attached his own letter of resignation in case the dean refused to accept of his recommendations. Geiling delivered the news of such decisions at the end of a breakfast at the University Club. John Doull recalled his experience at one meeting: "We discussed my going to medical school at one such breakfast and when I indicated that I would like to think about it, he suggested that I do so quickly since I was already enrolled in gross anatomy starting the following week."[11] The strength of Geiling's leadership drove the Department of Pharmacology and the Toxicity Laboratory forward, producing scores of papers and dozens of graduate students who would be central to the formation of the discipline of toxicology. Several important research programs distinguished the early years of the Tox Lab: the joint toxicity of antimalarial drugs, antimalarial drug resistance, nitrogen mustard compounds, and tracing minute doses with radioisotopes.

After its extensive effort to screen the toxicity of chemical warfare agents, the Department of Pharmacology under Geiling undertook an important research program to address the problem arising from the shortage of two antimalarial agents: quinine and atabrine. As a perennially deadly disease, malaria posed a great threat during times of war, especially in tropical regions.[12] The endemic and widespread presence of malaria, particularly in the Pacific theater, required the U.S. armed

UNIVERSITY OF CHICAGO TOXICITY LABORATORY

forces to find an effective therapy against it. DDT controlled malaria by reducing or eliminating mosquitoes, but infected soldiers needed immediate treatment after contracting the disease. During World War I, malaria had exacerbated the challenges of war, leaving troops debilitated and demoralized.[13] Malaria wreaked havoc among troops largely because medical personnel had received no special training on how to address the threat it posed. With the advent of World War II, officials recognized the danger of the disease and directed concerted efforts at its control, which incorporated staffs of the army, navy, the PHS, International Health Division of the Rockefeller Foundation, universities, and corporations, as well as the National Research Council and the OSRD.[14] Thus, on several levels, military public health officials sought to address the threats posed by malaria.

Certainly part of the reason for the increased awareness of malaria was the significant potential for exposure to the disease. American troops were deployed in Panama and the Caribbean, the west coast of Africa, the Balkans, Sicily and Southern Italy, India, Burma, Indonesia and the islands of the South Pacific, Formosa, and southern China, which were among of the most concentrated malarial regions of the world. Because medical entomologists had brought malaria under control in the United States during the early part of the twentieth century, few American soldiers had previous contact with the disease and virtually none had acquired immunity. Even those who had developed a level of immunity in the U.S. or elsewhere had not acquired immunity to the malaria strains found in Africa, Asia, and Europe.

Researchers in Germany, France, and the U.S. developed antimalarial therapies with widely variable results. For example, in the course of preparing and testing more than twelve thousand antimalarial compounds, Bayer, a division of I. G. Farben Industrie, which developed sulfanilamide (see chapter 1), produced in short order two new drugs: plasmochin (known as pamaquine in Great Britain) and quinacrine (mepacrine in Britain). U.S. researchers independently synthesized chloroquine and amodiaquin. In the early years of the war, American researchers studied several antimalarial drugs including atabrine and plasmochin.[15]

DDT significantly reduced the swarms of mosquitoes that carried the disease to human beings, but medical officials also sought to control the

77

UNIVERSITY OF CHICAGO TOXICITY LABORATORY

disease once it had invaded the human system. At the Tox Lab, Geiling and his collaborators tested existing drugs and searched for new ones, a program that proved highly successful not only in its immediate mission but also as a contribution to the growing methodology of environmental toxicology.

At the University of Chicago, Graham Chen in the Department of Pharmacology supervised the clinical investigations. Tox Lab scientists initially screened more than fourteen thousand drugs, of which approximately one hundred reached the clinical stage of investigation. Chicago became involved with the study of chloroquine after unforeseen toxicity developed in Marines treated with the drug by Coggeshall at Klamath Falls. This mishap prompted the OSRD to request that Chen and Geiling conduct toxicity studies on chloroquine at an Illinois prison. After some very quick arrangements, the Chicago researchers transferred their research from Manteno State Hospital to Stateville Penitentiary more than eighty miles away.[16] The toxicity studies of chloroquine at Stateville Hospital began on October 25, 1944. After several months spent establishing the facility, the Chicago toxicologists administered mosquito-induced malaria of the Chesson strain, Southwest Pacific Vivax, to thirty prison volunteers on March 8, 1945. This process demanded the better part of twenty-four hours and not all of the infections took hold. As the national malaria program turned from suppression of the disease to cure, the Chicago researchers quickly acquired a major portion of the available research funds.

Biomedical historian Nathaniel Comfort has shown that the Stateville malaria program occupies a precarious position in the history of biomedical ethics. On one hand, the Chicago researchers obtained what they understood to be informed consent from all the prisoners who "volunteered" for the malaria research. On the other, Comfort determined that many of the experiments conducted on prisoners could not have been performed on civilians in accordance with ethical standards at the time (or any time since). Yet Comfort argues that the Stateville malaria project defies simple ethical or moral formulas: prisoners willingly consented to "degrading, painful, dangerous, even life-threatening procedures" in spite of the risks.[17]

With the disbanding of the OSRD on July 1, 1946, funding for the project was transferred to the PHS. Less than eighteen months later, Chen, Geiling, and other researchers at Chicago had carefully screened

UNIVERSITY OF CHICAGO TOXICITY LABORATORY

thirty-five 8-aminoquinolines and had shown that isopentaquine was superior to pentaquine, which was better than pamaquin. Despite these promising results, the majority of the members of the Malaria Study Section at NIH believed that the examination of such compounds should be terminated.[18]

After submitting an application to NIH for a terminal grant in the amount of $15,000 on March 26, 1948, Chen and the University of Chicago team focused on the primary amines, which had been generally neglected. One of these compounds, which became known as primaquine, was one of the most promising discoveries of the antimalarial research at Chicago. Volunteers at the Stateville Penitentiary first received primaquine in the amount of 15 mg per day beginning on February 22, 1948. Researchers deliberately minimized the dose in response to earlier reports by Schmidt that primaquine was as toxic as pamaquin (later studies with monkeys refuted Schmidt's findings). On March 30, 1948 (only a few months before the termination of the grant), researchers raised the dose of primaquine to 30 mg per day. It was not until July 1, 1948, that Chen realized the considerable potential of primaquine. Given the imminent termination of research funding, Chen turned to Lowell Coggeshall, another malariologist, who was dean of the University of Chicago School of Medicine. Chen thought that he could run the antimalarial project at the most basic level (protecting infected volunteers without initiating new studies) or run the project at its fullest extent, an effort demanding $25,000 to $30,000 per year in order to explore primaquine. Coggeshall strongly encouraged Chen to explore the new drug: "Run Stateville full blast. If primaquine is as good as it seems to be, I do not believe you will have any difficulty finding sponsors for your work. If it turns out to be a false alarm, I will guarantee that University of Chicago will give you enough funds to protect you from trouble at Stateville."[19]

When the PHS announced that it was withdrawing support of all clinical investigations of malaria in the U.S. because malaria no longer constituted a public health threat, the army assumed financial support of the Stateville antimalarial study on January 1, 1950. Chen recommended against myopically limiting the study to primaquine, but in the course of subsequent studies, primaquine proved so effective against Korean malaria that development of new antimalarial drugs virtually ceased.[20]

UNIVERSITY OF CHICAGO TOXICITY LABORATORY

Before the incorporation of volunteers at Stateville Penitentiary, four of the Tox Lab scientists served as subjects for a study, a clear indication of their dedication to the research. Kelsey, Oldham, Dearborn, M. Silverman, and E. W. Lewis monitored the excretion of atabrine in urine. All of the "subjects" developed mild toxic symptoms in response to a daily dose for forty-five days, and the excreted amount of atabrine never amounted to more than 11 percent of a daily dose. Even fifty-five days after the last dose was administered, there were appreciable amounts of atabrine in the subjects' urine.[21] Studies involving self-experimentation suggest that the Tox Lab scientists were passionately committed to their research. At the same time, willingness to participate on the part of scientists may also indicate that they were not concerned about the toxicity of antimalarial drugs.

One of the most important studies that Chen and Geiling carried out under the OSRD grant assessed the joint toxicity of several antimalarial drugs.[22] With the considerable development of antimalarial drugs during and immediately following World War II, some physicians began to experiment with combinations of drugs with the expectation that they might be more effective than an individual drug in curing of disease. Chen and Geiling sought to determine the joint toxicity in the host as well as the efficacy of various combinations of several antimalarial drugs in mice. For guidance regarding dosage-mortality relationships, they turned, once again, to the research of Chester I. Bliss, who, as we have seen, had developed rigorous biostatistical approaches to dose-mortality curves and the LD_{50} for individual drugs. In addition, Bliss devised statistical methods for the evaluation of joint toxicities. The subject was of interest to Bliss on theoretical grounds: "What was the impact of multiple chemicals on a toxicity curve?" But Bliss also found joint toxicity interesting for practical reasons, particularly as it related to the search for new insecticides: "In the search for new insecticides combined poisons offer many possibilities, but criteria are needed for separating mixtures in which the combined ingredients possess an enhanced toxicity from others in which they act independently since the former group provides the more promising field of investigation." Bliss cited a study of the toxicity of two pesticides (rotenone and pyrethrin) used in combination in which the authors did not find evidence of synergism while another researcher utilized the same original data and discovered definite evidence of synergism.[23]

80

UNIVERSITY OF CHICAGO TOXICITY LABORATORY

In order to resolve such confusion, Bliss provided the definition and quantitative analysis of three kinds of joint toxic action in which the percentage mortality was employed as the measure of response. In the case of "independent joint action," the poisons or drugs acted independently and had different modes of mortality. Susceptibility of an organism to one component might or might not be correlated with susceptibility to the other. Quantitatively, the toxicity of the mixture could be predicted from the dosage-mortality curve for each constituent applied alone and the correlation in susceptibility to the two poisons. Bliss employed the term "similar joint action" for poisons or drugs that produced similar but independent effects, such that one component could be substituted at a constant proportion for the other. Individual susceptibility would be completely correlated or parallel. Quantitative calculation of the toxicity of compounds with similar joint action could be predicted directly from the toxicities of the constituents as long as their relative proportion was known. Finally, and perhaps most significant, Bliss delineated "synergistic action," in which the effectiveness or toxicity of a chemical mixture could not be assessed from that of the individual components but rather depended on knowledge of the chemicals' joint toxicity when used in different proportions. Synergistic action had the most serious implications for pharmacology and toxicology because one component exacerbated or diminished the effect of the other.[24] As Bliss predicted, his research and methodologies had wide application in the development and applications of drugs, insecticides, and other chemical mixtures.

For their part, Chen and Geiling directly applied Bliss's definitions and methods to antimalarial drugs. Atabrine and quinine, for example, acted in an independent and similar manner, as did quinine and hydroxy-ethylapocupreine. However, the combinations of quinine and pamaquine, as well as quinine and pentaquine, were much more toxic than predicted from their individual toxicities, clear cases of synergism in the two combinations. The dosage-mortality curves looked like those for different drugs rather than the summation of the curves for the individual drugs. Chen and Geiling explained the joint toxicity of atabrine and quinine by suggesting a common site of action, but they were at a loss to explain the synergism between quinine and pamaquine: "Since only a very small amount of quinine, 1/30 of the minimal lethal dose, is sufficient to reveal its synergistic action with a minimal lethal dose of

81

UNIVERSITY OF CHICAGO TOXICITY LABORATORY

pamaquine in acute mortality, the site and the mechanism of this action of quinine are evidently different from those causing sudden death of animals with a lethal dose of quinine."[25] Speculatively, Chen and Geiling suggested that the joint toxicity might result from an effect on an enzymatic process essential for life. Emphasizing acute toxicity, their paper mentioned chronic toxicity only in passing, cautioning that it may (or may not) correlate to acute toxicity.[26] This important distinction was often overlooked in the early toxicology and pharmacological literature. As we will see in chapter 4 and remaining chapters, joint toxicity became an issue of central importance in the study and legislation of pesticides.

Chen and Geiling also attempted to reveal the nature of drug resistance in a continuing study of drugs to treat trypanosomes (parasitic protozoans that cause trypanosomiasis), specifically "trypanocidal activity," in lab mice. Their first report on this research described a simple, quantitative method of assay of the therapeutic activity of antimonials and provided a comparison of the trypanocidal potency and the toxicity of well-known organic antimony preparations. They based the assay for the potency of trypanocidal substances on the suppression of infection and the cure of the disease. In a theoretical sense, the most important finding of Chen and Geiling's initial study of trypanosomes was the quantitative evaluation of the "therapeutic index," or the ratio of maximal tolerated dose to minimal curative dose. Earlier researchers developed the therapeutic index in a qualitative sense without regard to biological variation. Citing recent advances in quantitative pharmacology, Chen and Geiling adopted the 50 percent level (i.e., the ratio of maximal tolerated dose to minimal curative dose was two to one). They reasoned that the weight of an observation was greatest when the effect was 50 percent, which was highly desirable in an assay like the one they devised in which the number of animals was small.[27]

To address the problem of drug resistance, Chen and Geiling first developed an in vitro procedure for determining the antitrypanosome effect of antimonials.[28] With this procedure, a doctoral candidate (F. W. Schueler) joined Geiling and Chen, to complete the research necessary for his dissertation, and they developed a criterion of resistance based upon the inhibiting power of mapharsen (a standard drug of reference) on the glucose utilization by trypanosomes. The resistance factor was equal to the ratio of the 50 percent suppressive dose of mapharsen on

82

UNIVERSITY OF CHICAGO TOXICITY LABORATORY

glucose utilization for the resistant mouse strain to the 50 percent suppressive dose of mapharsen for the normal mouse strain. Schueler, Chen, and Geiling suggested that the determination of parasite resistance could be based on a criterion of 15 percent suppression for toxicity studies or 90 percent suppression for investigations of lethal dosages rather than on the 50 percent suppression level they used.[29] Thus the methodology could cover the full range of studies in toxicity.

The study of antimalarial drug therapies began as an attempt to replace and improve upon existing antimalarial drugs for the war effort. The war's end brought about changes in emphasis and funding, but Chen, Geiling, and others continued and expanded their research on antimalarials. In addition to expanding the constellation of drugs that controlled malaria, Chen and Geiling addressed more subtle aspects of the drugs, such as joint toxicity and resistance. Joint toxicity or potentiation and resistance became significant problems as toxicologists examined the toxicity of synthetic insecticides. Researchers drew upon techniques developed for antimalarial drug therapies in their subsequent research with pesticides.

In addition to studying antimalarial drug therapies, scientists at the Toxicity Laboratory at the University of Chicago devoted considerable effort to the analysis of nitrogen mustards.[30] In 1942, C. C. Lushbaugh, a pathologist in the laboratory, noted that mice gassed with nitrogen mustards had many fewer white blood cells than normal, and that the bone marrow and lymph nodes of the animals no longer formed blood cells. This discovery prompted Leon Jacobsen and Charles Spurr in the Department of Medicine at the University of Chicago to test the effectiveness of the nitrogen mustards against certain diseases such as leukemia, lymphosarcoma, and Hodgkin's disease.[31] This research resulted in some of the first chemotherapeutic agents against cancers.[32] One symptom associated with all these diseases is the presence of abnormally large numbers of white blood cells. The mustard compounds affected the blood cells much like X-rays. The research at the Tox Lab complemented research conducted in the Department of Pharmacology at Yale University (also sponsored by the OSRD), which revealed the potential of nitrogen mustards for use in chemotherapy.[33]

During the war, the Toxicity Laboratory was committed to the discovery and analysis of the most toxic compounds, either for potential

UNIVERSITY OF CHICAGO TOXICITY LABORATORY

use in combat or in anticipation of what the enemy might do. At the close of the war, however, members of the Tox Lab returned to some of the nitrogen mustard compounds they had deemed "relatively nontoxic" during the early part of the World War II. Some of these compounds affected white blood cells despite their lesser toxicity.[34] After the war, Tox Lab researchers discovered that certain nitrogen mustards attacked all proliferating normal tissues, and they belonged to a group of compounds known as "mitotic arrestors." Although the nitrogen mustards fell short of a cure for cancer, clinical trials revealed that they promoted extended remissions of the disease and reactivated the sensitivity of tumors to X-ray therapy. Both of these factors encouraged further research.[35]

Even as physicists from the University of Chicago and elsewhere raced to complete research on the first atomic bomb in 1944, scientists had begun to explore the potential of nuclear research to benefit society. The result of these musings was the development of the Jeffries Committee, chaired by Zay Jeffries. Other members of the committee included R. S. Mulliken (secretary), Enrico Fermi, James Franck, T. R. Harness, R. S. Stone, and C. A. Thomas. The Jeffries Committee convened to determine insofar as possible the future of a new field, which they called "nucleonics." In biology and medicine, the most promising avenue of research seemed to be the use of radioisotopes to examine basic problems in animal and plant metabolism, such as respiration, photosynthesis, fat and protein metabolism, and minor problems, for example, the role of micronutrients.[36]

In 1945, the CWS assumed the funding for the Tox Lab, but by the close of the war, its parent organization, the OSRD, had disbanded. The lab changed hands once again in 1947 and entered into a contract with the Atomic Energy Commission (AEC). The new source of funding attracted the interest of Toxicity Laboratory researchers to the new technology of radioisotope markers. Before the war, pharmacologists had to rely on chemical agents as tracers in pharmacological research. Such tracers included certain dyes or chemicals with specific properties.

Radioisotopes promised to greatly enhance and improve the ability of pharmacologists to trace the pathways of a given drug. Unlike chemical tracers, which diluted or otherwise affected the makeup of a drug, radioisotopes led to "natural tracers," suggesting that they shared iden-

84

UNIVERSITY OF CHICAGO TOXICITY LABORATORY

tical atomic structure with nonradioactive drugs. In fact, radioactive drugs were almost exactly like their nonradioactive counterparts except that some of the atoms in the labeled drug were radioactive and emitted radiation which scientists could track with the use of highly sensitive instruments, such as the Geiger counter, the scintillation counter, or the ionization chamber. With these devices Geiling and his colleagues could detect the presence of the smallest amounts of a drug and all its metabolized products in all organs and tissues of the body.

Geiling enumerated the numerous challenges posed by certain drugs that the use of radioisotopes in pharmacology could address: "A number of our most useful drugs have (1) complex chemical structures, (2) cannot as yet be readily prepared in the chemical laboratory, (3) are administered to patients in such small doses that, when distributed in the tissues and body fluids, the conventional biological and chemical methods are inadequate, (4) some of the available methods may be able to detect the unchanged drugs, but not the metabolites, (5) another important advantage of using labeled drugs is that they can be studied at a therapeutic or sub-toxic level."[37] Earlier drug distribution studies received the criticism that the doses used were well above the therapeutic level and at times even in the toxic range. The lack of sensitivity of tracer chemicals had required experimenters to use large doses.

With the cooperation of botanists, zoologists, organic chemists, and individuals trained in radioisotope techniques, Geiling and his colleagues developed methods to produce radioactive drugs using carbon-14. To do this, they grew plants like digitalis (*Digitalis purpurea*) and nicotine (*Nicotiana rustica*) in a closed system into which they introduced radioactive carbon dioxide (which the plants absorbed during the process of respiration). By drying and processing the digitalis one could obtain radioactive digitoxin; running nicotine through the same process produced radioactive nicotine. In both cases, Geiling was able to demonstrate a high degree of purity for the drugs. Even in the earliest, exploratory experiments with carbon-14, Geiling believed that the technique was far superior to other methods of marking: "This may be an advantage in the use of such materials in biological problems, since it permits the tracing of all carbon-containing metabolic fragments of the drug rather than only the single atom usually labeled in synthetic drugs."[38] Further experimentation with radioactive digitoxin revealed

85

UNIVERSITY OF CHICAGO TOXICITY LABORATORY

*E. M. K. Geiling and a colleague preparing radioactive
digitoxin. Courtesy of the University of Chicago Photographic
Archives apf1–06306, Special Collections Research Center,
University of Chicago Library.*

that the effect of the drug was not cumulative: "Our preliminary experiments indicate clearly that digitoxin cannot be regarded as a cumulative drug since it is largely metabolized and excreted in a relatively
short time. The mechanism of digitoxin action thus needs to be reap-
praised with the aid of these new techniques."[39] This further research
raised Geiling's expectations for the new technique in human experiments: "The use of radioactive digitoxin in suitable patients should
throw considerable light on the metabolic pathway of this important
agent."[40]

86

The use of radioisotopes as tracers was not limited to drugs synthesized from plants like digitalis and belladonna. A young doctoral student named John Doull (b. 1922) developed a method for the biosynthesis of radioactive bufagin as the basis for his doctoral thesis. Under the supervision of Geiling and Kenneth DuBois (a faculty member recruited during the war), Doull fed tropical toads (*Bufo marinus*) with radioactive algae supplied by W. F. Libby of the Institute for Nuclear Studies at the University of Chicago. Radioactive carbon-14 was distributed throughout the algae, which had been grown in a closed atmosphere containing radioactive carbon dioxide. As a preliminary test, Doull fed slugs on radioactive lettuce prepared according to Geiling's methods. Toads consumed the radioactive carbon-14 either in liver mixed with the algae or in the slugs. Doull went on to extract radioactive venom from the parotid glands of the toads, from which he then tried to isolate the bufagin by fractional crystallization. He acknowledged, however, that the effort required to extract venom of high radioactivity was much greater than to incorporate radioactive carbon dioxide into plant principles through photosynthesis.[41]

Having worked out the biosynthesis of radioactive digitoxin and other drugs, Geiling and the Tox Lab researchers employed the new technology to advance their knowledge of drug action. For example, radioactive digitoxin showed that the drug crossed the placental membrane in rats and guinea pigs. This finding had several grave implications for the use of drug therapies in pregnancy. First, embryonic tissue showed a marked ability to catabolize digitoxin (break it down into metabolites), or there may have been a selective penetration of the digitoxin metabolites across the placenta (indicated by the high metabolite-digitoxin ratio). Second, on a tissue-to-weight ratio, digitoxin and its metabolites appeared more concentrated in the embryonic heart than in the maternal heart.[42] The latter finding suggested that a given drug therapy for a mother would subject the developing fetus to a much higher dose of the drug. As we have seen in the case of joint toxicity and resistance, the differential effect of drugs and other chemicals on mothers and fetuses would become very important in the study of insecticides.

After several successful animal experiments with radioactive digitoxin, Geiling and his colleagues were prepared to use the new technology in tests with humans. In the first human experiments, Tox Lab scientists

UNIVERSITY OF CHICAGO TOXICITY LABORATORY

returned to the question of renal excretion. Their original study with dogs demonstrated that the animals excreted up to 46 percent of a single dose of the drug (a glycoside) in the urine. This finding sharply contradicted early experiments using only bioassay techniques without the benefit of radioisotopes. For the human tests, Tox Lab researchers administered radioactive digitoxin to three patients suffering from arteriosclerotic heart disease with congestive failure of varying degrees of severity. The subjects excreted radioactive digitoxin over a span of thirty-one to forty-two days, suggesting the drug's considerable level of persistence. Nevertheless, the level of digitoxin dropped off quickly during the first day (10 percent). By the end of the third day, each subject had excreted an additional 10 percent of the digitoxin. Between the seventh and eighth days, excretion leveled off. From a comparison between "unchanged" digitoxin and its metabolites, the researchers concluded that the major route of excretion of digitoxin in cardiac patients was through the kidneys.[43]

In a slightly larger study of eight patients with cardiac failure, Tox Lab researchers measured the disappearance of unchanged digitoxin from the blood and discovered two rate constants. One component showed a half-life of twenty-five to thirty minutes and another had a half-life of forty-eight to fifty-four hours. The first half-life may have represented the rate at which digitoxin in the blood was equilibrating with the various body tissues, and the slower rate could have represented the rate at which "loosely" bound glycoside was being liberated from the body tissues.[44]

Yet another study explored the metabolic fate of radioactive digitoxin. Using three terminal subjects, Tox Lab researchers administered multiple doses of biosynthetically labeled carbon-14 digitoxin intravenously. Tissue analysis revealed where digitoxin and its metabolites concentrated in the body. Most interestingly, the researchers found that the myocardium did not particularly attract digitoxin, whereas the kidney, gall bladder, jejunum, ileum, and colon all showed the highest concentration of unchanged digitoxin. Metabolites of digitoxin pooled in the gall bladder contents, jejunum contents, and spleen in the highest concentrations. The liver contained the largest amount of both digitoxin and its metabolites, suggesting to scientists that the liver was the major organ involved in the detoxification of the drug and confirming earlier

88

findings that the kidney was the major organ involved in the ultimate removal of digitoxin and its metabolic products.[45]

Clearly, radioactive digitoxin allowed a new level of sophistication in scientists' ability to monitor the metabolism of chemicals in living systems, including humans. In time, however, pharmacologists developed new methods to synthesize radioactive drugs, which resulted in much higher concentrations of radioisotopes. Chemically synthesized radioactive drugs facilitated experiments that were much more precise than those conducted by Geiling and Tox Lab researchers.[46] Nevertheless, the Tox Lab's research demonstrated the importance of radioactive technologies after World War II. Moreover, the spirit of the research, tracing minute doses within systems, became very important conceptually as toxicologists evaluated the risks posed by increasingly toxic insecticides at ever diminishing exposures.

In a related project after the Air Force took over the Tox Lab contract in 1951, John Doull along with Vivian Plzak and Mildred Root established a screening program for radio protective elements. For each chemical analyzed at the Tox Lab, the scientists determined the LD_{50} in male mice. This critical piece of information represented the only toxicological information for many of the agents analyzed. The resulting database of LD_{50}s was more valuable than the few radio protective elements identified.[47]

Under the parochial leadership of E. M. K. Geiling, the Toxicity Laboratory at the University of Chicago emerged as one of the leading centers for research in toxicology. Over the course of World War II, up to sixty scientists became employed in its toxicological research. Many more studied for graduate degrees and advanced training in the evolving discipline. The search for antimalarial drug therapies, the study of nitrogen mustards, research using radioisotopes in plants, and the search for radio protective elements yielded valuable data and new approaches and methodologies in the study of toxicology. Specifically, the antimalarial studies added to an understanding of the important toxicological concepts of persistence and synergism in chemicals as well as the critical theoretical issues of joint toxicity and drug/chemical resistance. This research transcended antimalarial drugs and also found application in pesticides. In the process of preliminary screening during the war, Tox

89

UNIVERSITY OF CHICAGO TOXICITY LABORATORY

John Doull inoculating a hibernating gopher with tagged digitoxin. Courtesy of the University of Chicago Photographic Archive apf1–05858. Special Collections Research Center, University of Chicago Library.

Lab scientists dismissed nitrogen mustards as "nontoxic," but their effect on white blood cells suggested promise as chemotherapeutic agents. In addition, radioisotopes facilitated tracking minute, subtherapeutic doses of a drug. In short, Geiling and the growing group of scientists at the Tox Lab explored a broad range of studies with implications for pharmacology and, more important, for toxicology as a distinct discipline. Moreover, such research clarified toxicological methodologies as well as the toxicity of numerous chemical agents.

CHAPTER 4

The Toxicity of
Organophosphate Chemicals

In the preceding chapters we have followed several episodes in the development of a notion of environmental risk. Along with other early cases, the Elixir Sulfanilamide tragedy refined scientific methodology with an analytical technique for deriving LD_{50}s and prompted passage of the Food, Drug, and Cosmetics Act of 1938. With the advent of World War II there was renewed interest in insecticides that could control the spread of malaria and other insect-borne diseases. DDT was the most promising of these, and its potential effects on target organisms, lab animals, wildlife, and humans underwent extensive analysis. Much of the interest in DDT was concentrated in governmental organizations—the PHS, the FWS, the USDA, and the U.S. armed forces. Such scrutiny demonstrated that DDT had opened a new era in insect control and toxicology. No other insecticide killed such a broad spectrum of insects without damaging the crops it was protecting. No other insecticide inspired such extensive investigation. For DDT, scientists extended the scope of toxicology to include effects on wildlife populations.

The wartime pursuit of an effective insecticide against malaria-carrying mosquitoes was just part of the fight against malaria (one of DDT's important uses). Scientists in the Toxicity Laboratory at the University of Chicago also sought antimalarial drug therapies and contributed new techniques to measure the joint toxicity of drugs and drug resistance. In addition to antimalarial drug therapies, Tox Lab scientists

91

TOXICITY OF ORGANOPHOSPHATE CHEMICALS

developed methods to trace minute quantities of drugs like bufagin and digitoxin by rendering them radioactive. Equally important, the Tox Lab laid the foundation for an independent discipline of toxicology by training graduate students and supporting research. Through his paternalistic direction, E. M. K. Geiling inspired these and other developments.

Despite the extensive publicity focused on it, DDT was only one of many insecticides that scientists developed during World War II.[1] Governmental organizations exhaustively tested DDT, but the task of evaluating other pesticides fell mainly to a young scientist named Kenneth DuBois (1917–73) who was working in the Tox Lab. Like many other scientists, DuBois joined the war effort shortly after completing his Ph.D. in physiology and biochemistry at the University of Wisconsin. At the Tox Lab, DuBois's research revealed several strong commitments. First, DuBois pursued the biochemical aspects of toxicology and stressed the importance of in vivo confirmation of the effects of toxic agents observed in in vitro enzyme studies. Second, he endeavored to develop methods to measure these effects quantitatively.[2]

Like DDT and other chlorinated hydrocarbons, the organic phosphate insecticides (later, "organophosphates" or "OPs") were first examined by German chemists as potential nerve gases to be used in combat.[3] Organic phosphate compounds link many phosphorous atoms to oxygen atoms (termed esters of polyphosphoric acids). Among the compounds investigated was diisopropyl fluorophosphate (DFP), which contained only one phosphorous atom. The Germans eventually discarded DFP as a nerve gas, but their experiments indicated that it inhibited cholinesterase, a critically important enzyme needed for the proper functioning of the nervous systems of humans, other vertebrates, and insects. It was the quality of cholinesterase inhibition that convinced physicians to use DFP to treat glaucoma, by reducing the abnormally high tension of the eyeball, and also myasthenia gravis, an autoimmune neuromuscular disease, for which it was more effective than treatment with eserine.[4] The first organic phosphate, HETP (hexaethyl tetra phosphate), emerged from Gerhard Schrader's laboratory at Farbenfabriken, Germany, in the early 1940s. Schrader discovered the insecticidal properties of the organic phosphates during the war, and the chemicals reached America in 1945, when the British and American Technical Intelligence Committee interrogated German chemists in the immediate aftermath of the

92

TOXICITY OF ORGANOPHOSPHATE CHEMICALS

*Kenneth Dubois, studying the effect of radioactive digitoxin on
beating mammalian heart. Courtesy of the University of
Chicago Photographic Archive apf1–05876, Special Collections
Research Center, University of Chicago Library.*

war. One of the first groups to gain access to the organic phosphates was
the Tox Lab, where DuBois and his associates recognized cholinergic
symptoms (i.e., changes in the action of the neurotransmitter acetylcholine) produced by the new chemicals and found that atropine would be
an effective antidote.[5]

In the U.S., there was considerable interest in the in the new insecticides because the organic phosphates could allegedly control aphids,
against which DDT was proving ineffective. The Tox Lab assumed the
responsibility for testing the toxicity of the new chemicals largely because

TOXICITY OF ORGANOPHOSPHATE CHEMICALS

the University of Chicago was near one of the major labs, Chemagro, where organic phosphates were synthesized. HETP ($C_{12}H_{30}P_4O_{13}$) was one of the organophosphorous chemicals that German chemists had developed, and under interrogation they compared this new compound to nicotine for its action in destroying aphids. Tox lab researchers could not locate any references to HETP's mechanism of action other than these possible nicotinic effects. Several researchers at the lab noted during routine testing, however, that animals showed symptoms similar to those produced by DFP. Such symptoms included muscular twitching, tonic and tonic-clonic convulsions, involuntary defecation, micturition (urination), and salivation. The case for parallel actions of the two chemicals was reinforced when researchers produced miosis by placing a dilute solution of HETP in the eyes of rabbits. In the case of HETP, the dilation effect lasted for five to twelve hours compared to three days in response to DFP. The similarities in the action of the two chemicals prompted DuBois to consider their possible effects on cholinesterase.

Through a series of in vitro and in vivo tests, DuBois and George Mangun (also a researcher at the Tox Lab and its director from 1946 to 1953) investigated the effect of HETP on cholinesterase. In the in vitro experiments, they measured HETP's effect on the cholinesterase of rat and cockroach tissue by adding solutions of the inhibitor dissolved in the buffer to the test system, which facilitated manometric measurement of the cholinesterase activity. The final concentration of 1×10^{-7} M HETP inhibited cholinesterase by 47 percent in the brain tissue of rats, by 53 percent in the submaxillary, 60 percent in serum, 45 percent in erythrocytes, and 58 percent in cockroach tissue. In comparisons of HETP with two recognized cholinesterase inhibitors (DFP and carbamic acid ester), DuBois and Mangun found HETP to be the most effective. For the in vivo experiments, the researchers administered HETP intraperitoneally (directly into the body peritoneum or body cavity) to rats and then measured the cholinesterase activity of the brain, submaxillary glands, and serum with the manometric test system. The results of the in vivo experiments verified the in vitro experiments, revealing cholinesterase inhibition in all of the tissues. Thus DuBois and Mangun concluded: "Hexaethyl tetraphosphate exerts a strong inhibitory effect on mammalian and insect cholinesterase in vitro and in vivo. This finding, in conjunction with its gross effects on animals, suggests that its physiological

TOXICITY OF ORGANOPHOSPHATE CHEMICALS

effects may be at least in part due to its inhibition of this enzyme."[6] This research connected the new insecticides with cholinesterase inhibition. The comparison with carbamic ester anticipated by nearly a decade the development of carbamate insecticides (see below).

DuBois and other researchers at the Tox Lab examined the toxicity of other organic phosphate insecticides as well. For much of this research, DuBois was joined by John Doull. For his doctoral dissertation, Doull used radioisotopes to evaluate cardiotoxic and other effects of bufagin. He received his Ph.D. in 1950 and his M.D. in 1953. Later Doull recalled that he had initially contributed to the analysis of the toxicity of organic phosphate chemicals.[7]

One of the most important chemicals investigated at the Tox Lab was a new insecticide called parathion. Parathion appeared to be particularly effective against plant insects, and its potential use stimulated researchers to examine its toxicity and pharmacologic action in mammals. Their approach to the analysis of parathion shared many similarities with the toxicological analyses of DDT. Along with Paul R. Salerno and Julius M. Coon, DuBois and Doull evaluated the acute and subacute toxicity of parathion as well as its inhibitory action on cholinesterase. They determined that LD_{50}s were low (less than 20 mg/kg) in all species (rats, mice, cats, and dogs), whether parathion was administered intraperitoneally or orally. Recall that a low LD_{50} corresponds to a very high toxicity. When sublethal doses of parathion were administered daily, its toxic action was cumulative. DuBois and his team also noted that the symptoms produced by parathion were similar in all species tested. These symptoms were typical of parasympathomimetic drugs (i.e., cholinergic drugs or those that mimic acetylcholine and inhibit cholinesterase). Like HETP, the Tox Lab researchers showed parathion to be a strong inhibitor of cholinesterase. In vitro a final concentration of 1.2×10^{-6} M inhibited by 50 percent rat brain cholinesterase. Finally, they explored possible antidotes to the lethal effects of the drug.[8] A picture of consistency within the class of organic phosphate insecticides gradually emerged from toxicological assessments like this one.

DuBois and his team also conducted the first toxicological evaluation of OMPA or Pestox III (Octamethyl Pyrophosphoramide). Like other organic phosphates, OMPA was first synthesized by Schrader, who demonstrated its insecticidal properties. He also noted that plants absorbed

95

TOXICITY OF ORGANOPHOSPHATE CHEMICALS

OMPA from the soil, which rendered them insecticidal. In fact, one group of researchers showed that OMPA had limited value as a contact insecticide, though plants grown in soil containing the chemical became highly toxic to insects for several weeks. This same group of researchers claimed that OMPA was toxic to mammals when mixed with food and administered orally.[9] On the whole, DuBois and his colleagues found the toxicity of OMPA similar to that for parathion. LD_{50} values for several species were virtually identical to those for parathion. All of the species exhibited symptoms typical of parasympathomimetic drugs except symptoms linked to the stimulation of the central nervous system. In regards to OMPA's potential as a systemic insecticide, DuBois's group noted that plants grown in soil containing OMPA contained an anticholinesterase agent. This finding indicated that plants converted OMPA much like the mammalian liver.[10]

In the Tox Lab, DuBois mainly studied the toxicity of organophosphates to animals, while other labs assessed the risks posed to humans from information gained through occupational accidents. David Grob and other researchers at the Johns Hopkins University School of Medicine reviewed the toxic effects of parathion in thirty-two men and eight women following accidental exposure. The research at Johns Hopkins, like that at the Tox Lab, was supported by the Medical Division of the Chemical Corps of the U.S. Army. Grob and his colleagues identified a disturbing characteristic of parathion: it could be readily absorbed through the skin, respiratory tract, conjunctivae, gastrointestinal tract, or following injection, most likely due to its high solubility in lipids (fats). Moreover, parathion did not produce inflammation in the skin so absorption could remain undetected.[11] That parathion could be absorbed through the skin contrasted with DDT, which had a low rate of dermal absorption, a real advantage in the eyes of economic entomologists. This basic difference between the two chemicals accounted for parathion's much higher level of toxicity. More troubling than dermal absorption was the absence of an inflammatory reaction, which suggested that an individual could suffer a toxic exposure without being aware of it. Still, the Johns Hopkins researchers were able to determine several general warning symptoms—intermittent nausea, vomiting, giddiness, weakness, drowsiness, and fasciculations (muscle twitches) of the eyelids—which appeared for one to seven days before the more severe manifestations developed.

96

TOXICITY OF ORGANOPHOSPHATE CHEMICALS

Grob and his colleagues also addressed symptoms that were particular to organic phosphate chemicals and specifically to parathion. They classified the symptoms with respect to two classes of action on the sympathetic nervous system: muscarine-like symptoms (anorexia and nausea, vomiting, abdominal cramps, excessive sweating, and salivation) and nicotine-like symptoms (nausea and vomiting, muscular fasciculations or twitches in the eyelids and tongue, followed by fasciculations in the muscles of the face and neck, in the extra-ocular muscles, and finally generalized fasciculations and weakness). Although categorizing symptoms may seem esoteric, the symptoms suggested that the organophosphates affected both the muscarinic receptors and the nicotinic receptors in the nervous system. The term "muscarine" tied the organophosphates to the long history of poisons. First isolated from a mushroom, *Amanita muscaria,* in 1869, muscarine was the first parasympathomimetic substance ever studied; it causes profound activation of the peripheral parasympathetic nervous system that may end in convulsions and death. More troubling still, the subjects developed all these symptoms and some of them died, despite the fact that most of them had worn carbon filter respirators, rubber gauntlets, and protective coveralls during their exposure to parathion. Some of the subjects had even worn hip-length rubber boots and rubber aprons. The Johns Hopkins researchers suspected that breaks in these safety procedures had occurred. Moreover, they discovered that the face piece respirators did not fully protect the workers from inhaling organophosphates as aerosols, dusts, or sprays.

Grob and his associates were able to isolate factors that accounted for the fact that parathion was more efficient than Tetraethyl Pyrophosphate (TEPP) as an insecticide even though its anticholinesterase activity and toxicity were lower. These factors also explained the greater danger of parathion to humans and domestic animals in comparison with other organic phosphates. The Johns Hopkins researchers noted that parathion hydrolyzed (broke down into chemical components) more slowly than TEPP, so that once sprayed on trees or fields, parathion remained active for weeks despite contact with moisture, whereas TEPP would hydrolyze within several hours. Another factor they noted was parathion's higher solubility in lipoid (fat), which meant that it would accumulate in the waxy outer layer of fruit and leaves, where it had been found up to nine days after spraying. Finally, the oxygen analogue of

97

TOXICITY OF ORGANOPHOSPHATE CHEMICALS

parathion was more toxic and active against cholinesterase than parathion, but Grob and his team could not determine the extent to which exposure to the air (as in spraying), or in plant tissues, or after absorption into the body converted parathion into its oxygen analogue.

By way of conclusion, the Hopkins researchers suggested a series of precautionary measures to be taken when dealing with parathion. Their recommendations included adequate warning labels, complete protective clothing, respirators, and a change of clothing before eating or smoking. They made specific recommendations regarding the use of parathion on crops for consumption: fruits and vegetables should be sprayed only with very dilute solutions, harvested no less than three weeks after the last spraying, and thoroughly washed prior to use.[12] Even within a class of highly toxic chemicals, parathion, as Grob and his team showed, was extremely hazardous to humans in the workplace and probably also as a food residue.

English physicians also examined the effects of poisoning by organophosphate chemicals. One of them, Lesley Bidstrup, reported only four organophosphate poisonings up to 1950, but each case stood out for the rapidity with which the insecticides wrought havoc on human systems. In one case, a plant foreman and fellow worker were splashed with parathion. Although the foreman made his assistant wash with soap and water and change his clothes at once, the foreman neglected to follow his own advice. After eight hours, he developed nausea, vomiting, abdominal cramps, diarrhea, and constriction of his pupils. By the time the foreman was admitted to the hospital nine hours later, he had developed fibrillary twitching of voluntary muscles and signs of pulmonary edema. Despite treatment with atropine, he died twenty-one hours after the accident.[13] With accounts such as this one in mind, Bidstrup bemoaned the lack of knowledge regarding long-term exposure to small amounts of parathion, although he suggested that if the need to leave sufficient time between spraying and harvesting were strictly observed, the food supply would remain safe. The potential exposure of workers was more worrisome, and Bidstrup concluded: "Experience in the United States of America in the 1949 spraying season has demonstrated that, unless all the recommendations made for the safe handling of organic phosphorus insecticides are carried out in detail, serious illness and even death will occur."[14]

Officials at the FDA also studied the new chemicals. From 1946 on, Arnold J. Lehman served as chief of the Division of Pharmacology at FDA. That year, a reorganization of the FDA established specialized sections. Under the new arrangement, the division of toxicology encompassed the acute, chronic, and dermal toxicity sections. Lehman assumed his post at FDA after a distinguished career in research and academia. Before joining the FDA, Lehman served as professor of pharmacology and director of the teaching and research activities in the Pharmacology Department at the University of North Carolina Medical School. In addition, he served as a consultant to the Federal Security Agency and as a member of the Committee on Atomic Research.[15] In June 1948 he presented a paper to the Association of Food and Drug Officials of the United States in Portland, Maine, titled "The Toxicology of the Newer Agricultural Chemicals." Lehman compared the toxicities of about two dozen insecticides including DDT, TEP (or TEPP), parathion, HETP, nicotine, chlordane, and heptachlor. Using DDT as a reference standard for insecticide toxicology, he listed the newer insecticides according to their acute oral toxicities. In this hierarchy, DDT had a median lethal dose of 250 mg/kg. In comparison, TEPP's LD_{50} was 2 mg/kg (or 125 times more toxic than DDT), parathion's LD_{50} was 3.5 mg/kg (70 times more toxic than DDT), and HETP's was 7 mg/kg (35 times more toxic than DDT). Lehman's hierarchy highlighted the relatively low acute toxicity of DDT.

Lehman also addressed three aspects of dermal toxicity: skin irritation, quantities dangerous upon skin application (single exposure and multiple exposure), and quantities dangerous to man (estimated). In a visually powerful manner, Lehman demonstrated that the organic phosphates were at least one order of magnitude more toxic than DDT. For example, Lehman estimated that it would take a single dermal exposure of 169 grams, and multiple exposures of 9 grams/day of DDT to be harmful to man, whereas for TEPP and parathion, he estimated single exposures of only 0.6 and 3 grams, respectively, of TEPP and parathion, and multiple exposures of 0.3 grams/day, were harmful. Despite their relatively higher dermal toxicity, the two organic phosphates irritated the skin only slightly (compared to no irritation for DDT). Thus an individual could suffer toxic exposure to an organic phosphate without noticing it.

TOXICITY OF ORGANOPHOSPHATE CHEMICALS

Lehman also described chronic toxicity in rats in his address, but an accompanying table revealed the paucity of data available from long-term studies. By 1948, few insecticides had undergone toxicity experiments lasting more than 52 weeks. DDT was one of the few insecticides that had been subjected to a two-year study of chronic toxicity to rats (see chapter 2). Thus Lehman could state that the lowest level of DDT producing gross effects was 100 parts per million (ppm), as demonstrated in a study lasting 104 weeks. For parathion, by way of contrast, Lehman listed 25 ppm as the lowest level producing gross effects. He based this claim on a study of only 4 weeks duration. Remarkably, even at levels of 1,000 ppm, HETP produced no effect over the course of a study lasting 12 weeks.[16]

Beyond the considerable value of his comparative tables, Lehman anticipated some of the most significant problems associated with the newer chemical insecticides. First, he undercut one of the fundamental beliefs behind the expanding use of pesticides: "It is a fairly safe assumption that chemicals which are toxic to insects are also toxic to man and animals. The great emphasis which has been placed on the specificity of DDT for insects loses its importance when fatal doses are compared on a body-weight basis with warm-blooded animals. On this basis the quantities required are practically identical."[17] Although he was extrapolating from limited data, Lehman's statement drew on his vast experience in pharmacology. He also expressed concern that the body stored certain insecticides like DDT in fat. Even more disturbing was the secretion of DDT and other chlorinated hydrocarbons in milk: "This is especially important in cases of infants, where the chief diet is milk."[18] Lehman's concern was not limited to DDT and the chlorinated hydrocarbons. Parathion was known to have a cumulative action, which pointed to its storage in tissues.[19] He reserved his most disturbing comment for the end of his paper: no one knew the dangers of using such chemicals in aerosol form. This information was available only for DDT, which had a safety factor several hundred times greater in such conditions. Lehman effectively outlined a comparison of the toxicology of the new agricultural chemicals and from this review identified some of the significant concerns regarding their widespread utilization. Moreover, he anticipated the litany of problems Rachel Carson attributed to pesticides in *Silent Spring* more than a decade later.

TOXICITY OF ORGANOPHOSPHATE CHEMICALS

In 1949 Lehman listed the insecticides in descending order of potential harmfulness to the public health, with emphasis placed on risks other than those related to the spray residue on foods. He arranged the toxicity of insecticides as follows: "TEPP > Parathion > Compound 497 > Nicotine > Compound 118 > Chlordane > Toxaphene > DDT > Rotenone."[20] On the important issue of spray residues (typically small amounts of pesticides that remained on foods), he noted that the values established as safe by the current experimental evidence were subject to change, and that they applied only to a single item of food:

Rotenone	5 parts per million
Pyrethrins	10 parts per million
TEPP	rapidly decomposed; known decomposition products not considered as a hazard
Parathion	2 parts per million
Gamma isomer	3–5 parts per million
DDT	less than 1 part per million if all of the food consumed is contaminated; 5 parts per million approaches the upper limit in any single item[21]

According to this table, the most toxic of the organic phosphates also decomposed the most rapidly into harmless, nontoxic products. So quickly would TEPP decay that Lehman and other scientists saw no need to set a residue level. Because of its slower rate of decay, the residue limit for parathion was 2 ppm. DDT, the subject of the most extensive scrutiny, received the lowest residue level. Moreover, Lehman cautioned that, because specific chemical methods for the isolation of many of the chlorinated hydrocarbon insecticides had not been developed, the detection of their presence in foods depended on generic organic chloride determinations. Thus food containing any detectable organic chloride residues should be regarded as contaminated.[22]

The most notable difference between the organic phosphates and other synthetic insecticides, DuBois and his colleagues at the University of Chicago had found, was that the organic phosphates inhibited cholinesterase in all species, including humans. All of the organic phosphates caused cholinesterase inhibition to some extent, but the new insecticides varied considerably in other aspects, such as persistence.

101

TOXICITY OF ORGANOPHOSPHATE CHEMICALS

The Committee on Pesticides of the Council on Pharmacy and Chemistry of the AMA reviewed the available information on the known organic phosphates in 1950. After a general description of three organic phosphates, DuBois and Grob, members of the committee, summarized their pharmacology and toxicity (both were recapitulations of earlier papers).[23] Additional committee members contributed to the review. For example, two doctors from American Cyanamid (one of the chief producers of organic phosphate chemicals) and another from the California Department of Health discussed clinical experience, briefly presenting eight fatal cases, which were mostly occupational exposures of various sorts resulting from lack of protective clothing, but included also a German biologist who attempted to determine the human tolerance for parathion through self-experimentation. Another tragic case involved a ten-year-old child who drank from a whiskey bottle containing TEPP and died in about fifteen minutes, before medical assistance became available.[24] Such accounts revealed the greater toxicities of most organic phosphates in comparison with the chlorinated hydrocarbons like DDT.

The rapid hydrolization of most of the organic phosphates appeared to reduce their risk in soil, but in their contribution to the committee's review, Lehman, Albert Hartzell, and J. C. Ward noted that the relatively slower rate of hydrolization of parathion posed a health hazard when it was used on turf. They also introduced evidence from animal studies: "Life-time feeding studies in rats at low dietary levels of parathion indicate no detectable cumulative effects below 25 parts per million. Animals fed levels above 25 parts per million and up to 100 parts per million, although they survived, displayed symptoms of nervous system poisoning and possessed an inhibition of blood cholinesterase in proportion to the increase of parathion over 25 parts per million in the diet."[25] Drawing on this information, Lehman extrapolated the risk to humans and recommended a safe residue level on any one item of the diet of approximately 2 parts per million of parathion.[26] Even as they proposed a safe residue level, Lehman, Hartzell, and Ward cautioned that only if parathion was applied strictly in accordance with the recommendations of the BEPQ of the USDA, with "particular reference to the time between the last spraying and the harvesting of the fruit," would normal weathering reduce parathion residues to this level of safety.

TOXICITY OF ORGANOPHOSPHATE CHEMICALS

Another question that Lehman and his collaborators raised was whether or not the peel of a fruit would be used in the preparation of a particular foodstuff. Even at the lowest effective spray concentrations, the peel of a fruit taken alone could carry a load of 2 to 3 ppm of parathion; this same concentration constituted 0.16 ppm extended to the entire fruit. This distinction was crucial: peeling the fruit before use, utilizing the whole fruit, or using the peel alone could change the level of exposure to parathion by an order of magnitude. In light of these variables, they concluded by underscoring the importance of adherence to recommended spray schedules: "If spray schedules recommended by qualified entomologists are followed, it is quite unlikely that a parathion spray residue problem will become serious."[27]

Given the composition of the AMA Committee on Pesticides—two industry doctors, two university toxicologists, and two government representatives from the USDA and the FDA—the conclusion probably reflects a compromise among committee members. As an FDA employee, Lehman may have realized that the political climate for the agency was less than favorable. For example, Clarence Cannon, who had shut down FDA's pesticide research in 1937, had become chairman of the House Appropriations Committee. From this vantage point, which he held from 1947 to 1964, Cannon and like-minded southern and midwestern conservatives wielded considerable influence over the federal government and especially the FDA.[28]

In an address before the Chicago Dietetic Association on March 15, 1950, DuBois took up the issue of food residues and food contamination by new insecticides, such as DDT and organic phosphates, as well as by the new systemic insecticides. He succinctly reviewed the state of knowledge in 1950 regarding the acute and chronic toxicity of each of the insecticides. From the practical standpoint of chronic toxicity, the chlorinated hydrocarbons had been a problem of major concern since their introduction: "The chlorinated hydrocarbons are stable toward hydrolysis, and spray residues may remain on fruits and vegetables for a long time. Continued ingestion of these contaminated products may thus produce a health hazard. Furthermore, these materials are fat-soluble, and the ingestion of contaminated forage by dairy cattle results in the appearance of insecticides in the milk where they are concentrated in the fat."[29] All of these factors associated with the chlorinated

103

TOXICITY OF ORGANOPHOSPHATE CHEMICALS

hydrocarbons contributed to the significant risk of chronic toxicity. Du-
Bois cited one of the few studies of chronic dietary exposure to DDT,
which showed that levels of 100 mg/kg DDT in food produced chronic
poisoning (in the form of liver damage) in rats during the two-year
study. Like Lehman, DuBois urged caution in the face of scientific
uncertainty, noting that chronic poisoning by these chemicals was a distinct possibility.[30]

In contrast to the chlorinated hydrocarbons, food contamination had
not been a problem with organic phosphate chemicals, such as HETP and
TEPP, because of their rapid hydrolysis when they came into contact with
moisture; spray residues on fruits and vegetables would lose their toxicity
before the foods were consumed. Even parathion, DuBois explained, although more stable toward hydrolysis than the other organic phosphates,
was rendered nontoxic before foods were harvested. But DuBois drew a
sharp distinction between typical organic phosphates and the systemic
insecticides, such as OMPA (organic phosphate chemicals applied to the
soil and taken up by the plants rendering the plants themselves insecticidal). What did this mean for possible food contamination? DuBois
noted that the insecticidal agent formed within plants from OMPA rapidly lost its toxicity rendering plants nontoxic to insects by the time the
plants reached maturity. DuBois wondered, however, about the potential
risk from plants harvested before they finished growing. Because those
plants could be dangerously contaminated, DuBois advised restraint in
the application of systemic insecticides, restricting use to non-food crops
or food crops that were never harvested before maturity.[31]

Thus, DuBois underscored the fundamental differences between
chlorinated hydrocarbons and organic phosphates. Chlorinated hydrocarbons like DDT did not cause acute poisoning after a single dose.
Research had demonstrated, however, that animals ingesting the new
insecticides for a long time could be poisoned. In contrast, acute toxicity posed the most significant risk with the organic phosphate insecticides, but their rapid hydrolysis greatly limited the threat of chronic
toxicity and food contamination. Finally, DuBois noted that organic
phosphates used as systemic insecticides presented greater risk than contact with organic phosphates because of their ability to be absorbed by
plants. DuBois's simple taxonomy of the risks associated with the three
new classes of insecticides captured their essential differences.

104

TOXICITY OF ORGANOPHOSPHATE CHEMICALS

Most of the research activity on organic phosphates discussed to this point was concentrated in two locations: the University of Chicago Toxicity Laboratory under the supervision of DuBois and the FDA Division of Pharmacology with Lehman as its chief. Additional contributions to the literature of organic phosphates came from Grob at Johns Hopkins. Both DuBois and Lehman presented hierarchies of the risks posed by the various new insecticides within the broad categories of chlorinated hydrocarbons, organic phosphates, and systemic insecticides. Although the various groups seemed to work independently from each other, in fact, DuBois, Doull, and other researchers at the Tox Lab worked closely with Dan MacDougall and Dallas Nelson, the scientific staff at Chemagro (later Bayer Corporation) in Kansas City to plan and execute studies and eventually to defend new insecticides before the FDA. John Doull recalled interactions with the FDA: "These meetings were usually held in the FDA commissioner's office with Drs. Arnold Lehman, Garth Fitzhugh, Bert Vos and Arthur Nelson representing FDA and DuBois, Doull and MacDougall representing Chemagro. In contrast to the complex and lengthy procedure currently required to obtain pesticide tolerances, these meetings were short, informal and focused on the science (toxicology and pathology) rather than on any of the legal or political considerations that often seem to be of primary importance today."[32] It was not until Rachel Carson published *Silent Spring* in 1962 that the political nature of pesticide regulation came to the forefront of attention among the wider American public.

In 1952, DuBois and Julius M. Coon, a doctor in the Tox Lab, returned to the toxicology of organic phosphorous-containing insecticides to mammals. DuBois and Coon classified the organic phosphates into three groups based on the chemical formula of each insecticide: alkyl pyrophosphates, alkyl thiophosphates, and phosphoramides. Among the alkyl pyrophosphates, TEPP was the most important, and DuBois and Coon reconfirmed the considerable toxicology of TEPP and particularly cholinesterase inhibition. In an analysis of additional alkyl pyrophosphates, DuBois and Coon demonstrated that they all exhibited cholinergic properties similar to TEPP. Several important organic phosphates, including parathion, malathon, and systox, were classified as alkyl thiophosphates. In light of the extensive use of parathion as an agricultural insecticide, DuBois and Coon reviewed the akyl thiophosphates to find a compound

105

TOXICITY OF ORGANOPHOSPHATE CHEMICALS

as toxic as parathion to insects but less toxic to mammals. They showed that the LD_{50} for rats for parathion was 5.5 mg/kg while that for malathon (later, malathion) was much higher, at 750 mg/kg. This was one of the first references to the toxicity of this newly developed insecticide. DuBois and Coon urged that these results be interpreted cautiously, since chemicals with a low toxicity for mammals generally exhibited a lower toxicity for insects and thus required use of higher concentrations in the formulations used in insect control. An ideal compound would have a high toxicity for insects and a low mammalian toxicity.[33] Thus, the two scientists pointed to one of the paradoxes of insecticide development. Insecticides of a lower toxicity to mammals often necessitated higher concentrations or quantities to produce the same measure of insect control. Raising the concentration or quantity undermined the advantage in toxicity.

In their consideration of the phosphoramides, DuBois and Coon shed light on one such biochemical interaction. They pointed out pharmacologic properties of OMPA (the only phosphoramide released for use at the time) that were unusual among the organic phosphates: "It exhibits no appreciable anticholinesterase action *in vitro* but is converted by the mammalian liver and by plants into a strong cholinesterase inhibitor. A further differentiating feature is its inability to gain access to the brain *in vivo*, its cholinergic action being therefore limited to peripheral tissues."[34] Of the organic phosphates, only OMPA, according to DuBois's experience, could be converted by the mammalian liver and plants into a cholinesterase inhibitor.

Their reference to malathon indicates that in the Tox Lab DuBois and Coon had access to the newest insecticides, even those that were still in development stages. The first complete review of the toxicity of malathon did not appear until 1953, when Lloyd W. Hazleton and Emily G. Holland of the Hazleton Laboratories in Falls Church, Virginia, summarized mammalian investigations of the new chemical. In collaboration with the American Cyanamid Company, Hazleton Laboratories selected malathon from a coordinated screening program. From entomological data, Hazleton and Holland believed that malathon would find wide use as an insecticide and that this might lead to appreciable human exposure. Because their preliminary data suggested considerable variation between insect and mammalian toxicity, they conducted fur-

106

TOXICITY OF ORGANOPHOSPHATE CHEMICALS

ther experiments on the acute toxicity of the substance to several different kinds of animals: "Regardless of technical grade, solvent, species,
sex or route of administration, the acute signs of toxicity are characteristic of the anticholinesterase activity. In rats, mice, guinea pigs, and dogs,
salivation, depression, and tremors predominate. The signs are of short
duration, and *unless death occurs within a few hours recovery appears to be
complete.* This observation should be emphasized, for later studies indicate that cholinesterase inhibition endures far beyond any gross evidence
of toxicity."[35] This statement suggests the possibility of a threshold for
the effects of cholinesterase inhibition. Above a certain level of exposure, laboratory animals died. Below that level, Hazleton and Holland
claimed, animals recovered completely from the exposure. The ability
of animals to recover from the anticholinesterase activity of malathon
served as a theme of their research.

Hazleton and Holland injected various concentrations of the new
chemical into guinea pigs, dogs, and albino rats and monitored the animals to determine the exposure levels that produced cholinesterase inhibition of 50 percent (Inhibition$_{50}$ or IN$_{50}$) in the animals' red blood
cells, plasma, and brain. As one example, the Hazleton Laboratory researchers subjected rats to intraperitoneal dosages of malathon, which
varied from 50 to 500 mg/kg. The IN$_{50}$ for red blood cells was 480 mg/kg
and 500 mg/kg for the brain. They compared these figures to the IN$_{50}$'s
for parathion (determined after 1.5 hours): 1.65 mg/kg for red blood
cells and 3 mg/kg for brain. According to these experiments, malathon
was at least two orders of magnitude less toxic than parathion. In
chronic feeding experiments conducted on rats, Hazleton and Holland
found no evidence of cholinesterase inhibition at 100 ppm malathon in
the diet, but it did inhibit cholinesterase by 73 percent in red blood cells
at 1,000 ppm/day and 100 percent at 5,000 ppm/day. In a two-year
feeding experiment, rats on a daily diet of 100 ppm malathon showed
slight evidence of cholinesterase inhibition. The results of the experiments conducted at the Hazleton Laboratories certainly demonstrated
not only that malathon was much less toxic than parathion, but that it
seemed to be the least toxic of all the organophosphate insecticides.

Hazleton and Holland used their results with malathon to challenge
DuBois and Coon's opinion regarding the organophosphate insecticides: "These data suggest that it would be timely to reconsider the view

107

expressed by DuBois and Coon that those materials which have a low toxicity for mammals generally exhibit a low toxicity for insects."[36] Hazleton and Holland argued that while parathion was approximately 100 times as potent in vitro and 135 times as toxic to rats as malathon, "Under usage conditions, no more than two to three times as much malathon as parathion is recommended."[37] Hazleton and Holland believed that they had discovered an insecticide that was highly toxic to insects but minimally toxic to mammals. To determine whether or not that was the case required significantly more experimentation on both target and nontarget organisms. Would malathon control insects effectively at nontoxic levels? Hazleton and Holland harbored even greater hopes for the new chemical. Beyond its specific value as an insecticide, they expected it to transform thinking about insecticide toxicity: "It is to be hoped that this compound will serve to point the way toward a better understanding of the difference between mammalian and insect toxicity and to free our thinking from the dogma that anticholinesterase activity in vitro is necessarily an index to mammalian toxicity."[38] It is tempting to conclude that these findings seem tainted by the fact that American Cyanamid funded research at the Hazleton Laboratory or that the Hazleton Laboratories researchers failed to place a convincing distance between their objective findings and their chief source of support. The findings at Hazleton Labs suggested that toxicity of malathion constituted an exception to the rule that placed most organophosphates among the most toxic chemicals known to mankind.

Two years later, Kenneth DuBois returned to the subject of malathon (renamed malathion in 1953). With Robert Bagdon, DuBois examined the pharmacologic effects of chlorthion, malathion, and tetrapropyl dithionopyrophosphate in mammals. Bagdon and DuBois cited DuBois's earlier work on the low toxicity of these compounds as well as Hazelton and Holland's determination of the low toxicity of malathion. For this study, they considered the effects of thionophospates on blood pressure, respiration, the isolated heart, and the intestine in vitro and in vivo. Bagdon and DuBois concluded: "On the basis of toxicity and associated pharmacologic effects the newer thionophosphates employed for this investigation possess a distinct advantage over others such as parathion and systox from the standpoint of the dose required to produce acute poisoning. Hence, the possibility of accidental poisoning during handling is considerably less than with agents such as parathion."[39] This statement was cautiously couched in terms of toxicology and pharmacology (DuBois's expertise), but it does not address the other part of the equation: would these insecticides necessitate greater quantities to affect the same control of target insect populations? DuBois restricted his conclusions to his area of specialization (mammalian toxicity). He and Bagdon did take up the issue of purity, however. They acknowledged that toxicity rose with impurity or contamination and cited Hazelton and Holland's finding that malathion became less toxic with increasing purity. Although Bagdon and DuBois clarified the toxicity of thionophosphates, including malathion, they left open the question of the quantity required to achieve an effectiveness equivalent to that of a more toxic cholinesterase inhibitor like parathion.

In addition to malathion, chemical companies developed other insecticides. Carbamate insecticides were a promising new class of pesticides. Union Carbide developed and released carbaryl or Sevin in 1956. Like organophosphates, carbamates inhibited cholinesterase. However, researchers at the Mellon Institute in Pittsburgh found that Sevin's anticholinesterase activity was greater against insects than against mammals. Tests with cats, guinea pigs, rats, rabbits, and chickens revealed LD_{50}s for various routes of exposure (oral, intravenous, intraperitoneal, and subcutaneous) in the range of 125 mg/kg in cats to greater than 500 mg/ kg in rats and rabbits. Two-year chronic feeding studies showed that rats tolerated daily doses of Sevin at levels up to 200 ppm. Similar studies demonstrated that dogs tolerated up to 400 ppm of Sevin in their diets on daily basis.[40] Thus, like malathion, the toxicity of Sevin to mammals was relatively low. Early researchers also noted that stability, anticholinesterase activity, and insect toxicity were different for the organophosphates and the carbamates.[41]

As additional organophosphates and other chemicals entered the public market during the 1950s, DuBois and his research team at the University of Chicago continued to evaluate their toxicity. Among the chemicals that they evaluated were Systox, Di-Syston, and other organophosphates. At the close of the decade, DuBois sought to extend the implications of more than a dozen years of research on the organic phosphate insecticides. Together with his student Sheldon Murphy, who had become a University of Chicago Fellow, he assessed the influence of

TOXICITY OF ORGANOPHOSPHATE CHEMICALS

various factors on the enzymatic conversion of organic thiophosphates to anticholinesterase agents. Their research transcended the limits of research on organic phosphates and merited further study. They concluded: "The results of the present investigation have provided some information on the mechanisms responsible for age and sex differences and other factors which influence susceptibility to cholinergic thiophosphates. The findings suggest that further research along similar lines may aid in gaining an understanding of the reasons for age, sex, species and individual differences in susceptibility to drugs and other chemical agents which have been observed frequently but have not been adequately explained."[42] They found that the enzyme activity of the livers of adult male rats was two to three times greater than that of adult females of the same age. There were no observable differences between the sexes of animals less than thirty days old. Yet Murphy and DuBois noted a dramatic increase in the liver activity of male rats between thirty and sixty days of age. This time period corresponded with the age of puberty.

The Chicago researchers next increased the low enzyme activity in adult females and young males by administering testosterone for a prolonged period. They also reduced the high enzyme activity in adult males by castration and through the extended administration of progesterone and diethylstilbestrol. These experiments indicated that sex hormones influenced the synthesis of the thiophosphate-oxidizing enzyme. Equally important, Murphy and DuBois determined the role of diet and nutrition in enzymatic activity: "Feeding a protein-free diet to adult male rats reduced the ability of the liver to convert guthion to an anticholinesterase agent by 75 percent. The increase in activity of the thiophosphate-oxidizing enzyme which occurs after the administration of carcinogenic hydrocarbons was inhibited when the animals were fed a protein-free diet."[43] By enabling the liver to convert guthion, dietary protein contributed to cholinesterase inhibition.[44]

Research on the toxicity of the organophosphate insecticides also continued at the FDA. In many respects, the FDA research complemented DuBois's numerous studies on the toxicity of the organophosphates. Scientists at the FDA endeavored to develop and test methodologies for the analysis of the toxicology of the new synthetic pesticides. J. William Cook was a biochemist in the Division of Food at the FDA from 1951 until 1972. He also served as the director of the Division of Pesticide

110

TOXICITY OF ORGANOPHOSPHATE CHEMICALS

Chemistry and Toxicology.[45] Cook searched for ways to employ the enzyme systems he had devised in his previous position in the San Francisco regional office. It occurred to him that one of the best applications of his enzyme research would be the analysis of organophosphate compounds because they were toxic by virtue of the fact that they inhibited cholinesterase. Cook explained the nature of cholinesterase inhibition: "The cholinesterase enzymes hydrolyze to a compound called acetylcholine. Acetylcoline is involved in the transmission of nerve impulses. Therefore muscle activity is based on acetylcholine being formed and hydrolyzed quickly. When those enzymes are inhibited, the person becomes rigid or has tremors."[46]

As Cook surveyed the literature on the esterase systems and their inhibition by the organophosphates, he discovered that such research required relatively complicated, expensive pieces of equipment, which he knew the FDA, with a total budget of five million dollars for all of its programs, could not afford to acquire. It was clear to Cook that budgetary restrictions put the FDA at a real disadvantage in its efforts to regulate pesticides when chemical companies had much more sophisticated analytical equipment. But fiscal constraints at the FDA inspired methodological creativity. Having developed a test for urea by putting the enzyme urease in paper along with a dye that would change with acid-base, Cook considered how he might develop a similar test for the organic phosphates—but the enzyme spot test was not his first thought. Initially he analyzed the organic phosphates using chromatography.[47] Learning from a colleague that the sulfur in most organic phosphates might be sensitive to bromine, Cook developed a spot test for organic phosphates. As in the urea test, he sprayed a paper with a bromine-containing compound and superimposed a dye chemical. Wherever the sulfur in the organophosphate used up the bromine, it was not available to change the color of the dye.[48] Moreover, the bromination technique converted the non-cholinesterase in vitro inhibitors to in vitro inhibitors of cholinesterase. With this knowledge, Cook could visualize some of the general chemical characteristics of these compounds. Using this approach, he learned to look for many useful signs when petitions came in for new organic phosphate compounds. He was thus able to accept or reject the data that companies submitted in their petitions based on the bromination technique.[49]

111

TOXICITY OF ORGANOPHOSPHATE CHEMICALS

Cook combined the two tests (the anticholinesterase method of analysis with paper chromatography and his newly devised brominated spot test technique) to analyze numerous organic phosphate chemicals, including parathion. The literature indicated that parathion was highly toxic to dogs (exposures as low as 1 ppm depressed cholinesterase). In contrast, large quantities fed to cows did not inhibit cholinesterase or cause it to appear in the cow's milk. Cook believed that something was happening to the parathion before it reached the bloodstream of the cow because in most mammals parathion fed at toxic levels moved from the bloodstream into the milk and meat. Cook fed parathion to a cow with an opening in its rumen (where he assumed the cow would break down the parathion). By the time he returned to his laboratory, the parathion had disappeared from the samples. In a review of the literature, Cook discovered that parathion had been reduced to a far less toxic amino group. From the combination of his own experimental data and his review of the literature, Cook felt confident that he could approve the use of parathion on plants fed to dairy cows because he knew it would not be transferred to their milk.[50]

In addition to the paper chromatography test and the test for anticholinesterase activity, Cook and a colleague, D. F. McCaulley, resurrected Edward Laug's fly bioassay (see chapter 2) for determining organic phosphate pesticides. Other researchers had developed bioassays using flies, but most of them were based upon mortality induced by graded amounts of pesticides. Such bioassays were very sensitive but unspecific. McCaulley and Cook felt that by linking a measurement of in vivo depression of fly cholinesterase to a fly mortality count, group specificity might be added to the assay's sensitivity. This procedure demonstrated the presence of any chemicals from the group of phosphate pesticides with legal tolerances for food residues (parathion, systox, methyl parathion, guthion, phosdrin, trithion, diazinon, malathion, and OMPA). The fly bioassay was effective as a screening procedure. Those samples showing significant mortality could be checked later for cholinesterase inhibition. Cholinesterase inhibition roughly equal to mortality indicated a phosphate as the main toxic factor; a mortality figure much higher than inhibition indicated the presence of a combination of toxicant, not all phosphate; and the absence of inhibition in the presence of considerable mortality would reveal a toxic factor other than a phosphate.[51] A

TOXICITY OF ORGANOPHOSPHATE CHEMICALS

technique that had been abandoned in favor of chemical methods was revived and effectively redeployed for use in a new context (detection of pesticide residues on foods). Refining such techniques ultimately led to the development and publication of the *Pesticide Analytical Manual,* which became the standard reference for testing the toxicity of pesticides.

Another concern for toxicologists was the toxicity of insecticides used in combination, or potentiation. A team of FDA pharmacologists led by John P. Frawley analyzed the greater than additive toxicity, or potentiation, resulting from simultaneous administration of two anticholinesterase compounds, which was essentially a study of joint toxicity. After reviewing the rather sparse literature on the toxicity of organophosphate insecticides, Frawley and his colleagues noted: "In all these studies, the observations have been based on the continued administration of a single compound. In practice a worker may be exposed to two or more compounds on the same or alternating days, and the average consumer may ingest at the same meal several different food products each containing a different insecticide."[52] Unlike the joint toxicity of antimalarial drugs as studied at the Toxicity Laboratory, which would have to be prescribed by a physician, individuals could be exposed to two organophosphates inadvertently through occupational exposure and possibly even normal daily consumption.

Frawley and his team chose two organophosphates, EPN and malathion, because they were each less toxic than other organophosphates. First they determined the acute toxicity (LD_{50}) of each chemical for rats and dogs, and then they established the toxicity of the two chemicals in combination. In dogs, EPN and malathion administered simultaneously caused up to fifty times the potentiation (additive toxic effects) of separate exposures. And they noted potentiation in rats as well. From these findings, Frawley and his team concluded: "However, of broader significance is the conclusion that in some cases the hazard associated with the administration of chemical and drug combinations cannot be evaluated from the toxicity of the individual compounds. The results point out the need for caution in the use of drug combinations in this phase of pharmacology and toxicology which is frequently overlooked."[53] The FDA group also investigated the joint toxicity of malathion and EPN combined in several ratios to house flies, again using Laug's fly bioassay, but

113

TOXICITY OF ORGANOPHOSPHATE CHEMICALS

found no indication of potentiation. This finding suggested that potentiation involved complex chemical reactions between the two phosphates and the biological system.

At the Tox Lab, DuBois also addressed the potentiation of organophosphates. He reasoned that the simplest method for detecting potentiation by acute toxicity tests would be to administer half of the LD_{50} of each of two organic phosphates. If mortality due to the combination of the two compounds was additive (50 percent) or less than additive, no potentiation had occurred. DuBois used this approach to test for potentiation in several organic phosphates and found that most showed additive or less than additive acute toxic effects. This meant that the combination of half of the LD_{50} of the two chemicals produced a toxic effect that was equal to or less than the full LD_{50} dose for either chemical. DuBois anticipated these results when the compounds had the same mode of action, parallel dosage-mortality responses, and a similar time of onset of toxic effects. From the results of the tests of acute toxicity, it became clear to DuBois that he had to clarify the mechanism of toxicity for each organic phosphate involved in potentiation to fully explore subacute effects. Such research revealed that some agents inhibited hydrolytic detoxification reactions. DuBois thought this discovery was potentially valuable for basic research into normal metabolism, but it left unresolved the implications for food residues and occupational exposures.[54] He noted, "Our present knowledge of the problem of potentiation of the toxicity of organophosphates does not provide an answer to the question of whether or not this effect constitutes a health hazard in connection with consumption of contaminated food."[55]

The organophosphate insecticides, like DDT and other chlorinated hydrocarbons, demanded novel toxicological techniques and strategies. During World War II, the Toxicity Laboratory at the University of Chicago responded to this considerable need. In particular, Kenneth DuBois and his students and colleagues recognized the major toxicological effects of the organophosphates: cholinesterase inhibition. DuBois and his research group developed toxicological profiles for many of the new insecticides, including OMPA and parathion. In addition to the research conducted at the Tox Lab, David Grob at Johns Hopkins examined the toxicity of parathion to humans, drawing on occupational cases of exposure. Arnold Lehman at the FDA also evaluated the risks of the organo-

114

TOXICITY OF ORGANOPHOSPHATE CHEMICALS

phosphates, particularly in comparison to other insecticides like the chlorinated hydrocarbons. Like DuBois, Lehman constructed hierarchies of toxicity for the new chemicals. In general, the organophosphate insecticides had a greater acute toxicity (due to cholinesterase inhibition) but considerably reduced chronic toxicity in comparison with the chlorinated hydrocarbons. One promising exception to this developing rule was malathion, or so American Cyanamid and scientists associated with the company argued. In the mid-1950s, Union Carbide introduced the first of yet another promising class of insecticides: Sevin. The carbamates inhibited cholinesterase like the organophosphates. Like malathion, Sevin had a relatively low mammalian toxicity. Combination of certain organophosphates exacerbated their effects as researchers at the Tox Lab and the FDA independently discovered. As toxicologists at the Tox Lab and the FDA strove to assess the risks associated with exposures to the new organophosphates, legislators held hearings to determine the implications for public health.

CHAPTER 5

What's the Risk?

Legislators and Scientists Evaluate Pesticides

In the aftermath of World War II, when DDT and other chemical insecticides became widely available for use in the U.S. for agriculture and public health, legislators began to realize the limits of the Federal Food, Drug, and Cosmetic Act of 1938 to regulate novel synthetic insecticides. Congress held several hearings to discuss further legislation. Industry representatives bridled at the idea of further regulation, and scientific opinion regarding risks of insecticides varied widely. But chlorinated hydrocarbons like DDT and the organophosphate insecticides taxed the regulatory power of the FFDCA, and Congress revisited the mounting challenges of synthetic insecticides by holding hearings, which led to the Federal Insecticide, Fungicide, and Rodenticide Act (FIFRA) in 1947. Despite wide-ranging hearings and the passage of the FIFRA, lawmakers returned to the subject of pesticides again in 1951 in hearings regarding food additives, with further legislation in the form of the Miller Amendment and the Delaney Clause.

The FIFRA hearings brought together representatives from government agencies (USDA and FDA), industry (the Agricultural Insecticide and Fungicide Association), and, to a limited extent, consumers. Among those who appeared before the congressional hearing was S. R. Newell, then assistant director, Livestock Branch, Production and Marketing Administration, USDA, who characterized the bill as follows: "The broad objective of this bill is to protect the users of economic poisons by re-

116

quiring that full and accurate information be provided as to the contents and directions for use and, in the case of poisons toxic to man, a statement of antidote for the poisons contained therein. It is also designed to protect the reputable manufacturer or distributor from those few opportunists who would discredit the industry by attempting to capitalize on situations by false claims for useless or dangerous products."[1] It should be noted that the "users of economic poisons" referenced here were farmers not consumers or the public at large. Newell noted unanimous agreement on the need for such a bill and general agreement that the Insecticide Act of 1910 no longer met the needs for effective regulation, largely as a result of the emergence of new pests, new insecticides, and new controls that had emerged over the course of the previous thirty-five years. New insecticides, such as DDT, inspired questions. It is worth noting, however, that the Insecticide Act of 1910, like the Pure Food and Drug Act, served primarily as a labeling law (see chapter 1). Newell suggested that the insecticides industry would react favorably to registration: "Experience over many years indicates that many manufacturers would welcome the opportunity to check their products and the claims made for them with the administrative body. Recent experience in the examination of labels applying to DDT amply demonstrates this fact."[2]

Newell's optimism regarding industry cooperation with pesticide registration was not shared by all. L. S. Hitchner, executive secretary of the Agricultural Insecticide and Fungicide Association (the national pesticide trade organization), stated his preference for free market competition over federal regulation: "Let us take DDT. I do not think the Bureau of Entomology or any government department or agency today is in position to set and freeze standards. Normal competition has given the American farmer the highest quality of goods in a highly competitive industry, and setting of standards, in my opinion, would be impossible."[3] Note that like Newell, Hitchner made specific reference to DDT. But Hitchner restricted his statement to the "quality" of insecticides, which presumably referred to their efficacy against target organisms rather than their potential toxicity to nontarget organisms, including humans. Moreover, Hitchner questioned the ability of government or industry to standardize insecticides, though he acknowledged that two older insecticides, arsenate of lead and calcium arsenate, were in fact standardized.

WHAT'S THE RISK?

He argued that state experimental stations could educate consumers in the use of pesticides, while dismissing standards: "For example, New York State today is having dealer conferences all over New York to educate buyers on insecticides. There is no simple way of arriving at a standard on two or three thousand chemicals. I wish we could, but it just seems to me to be impossible."[4] Hitchner argued that education provided by the state experimental stations obviated the need for federal standards, which would be impossible to develop anyway.

Hitchner returned the theme of state sovereignty later in his testimony when he challenged the consolidation of regulatory authority in the office of the secretary of agriculture by citing slow acceptance of oil sprays: "When oil sprays were first introduced, they were vigorously opposed by several of the state agricultural colleges and official workers of the government. The companies that introduced those hired their own entomologists, their own pathologists, and went from farm to farm getting the material used over the vigorous objection of the experiment-station people, who were in a rut on new development. Today there are millions of gallons of those oil sprays in commercial operation."[5] Once again, Hitchner assumed that reticence to adopt oil sprays stemmed from fear of new technology rather than safety concerns. But concentration of authority troubled the industry representative most: "You are definitely giving a man a right to say 'You cannot use cryolite; you have to use arsenate of lead'; or 'You can't use arsenate of lead, you have to use DDT.'" He continued: "The best example on that is where we made a survey on DDT, where we got 48 States to send their directions for use, and if you read the 48 reports there is hardly one State that agrees with any other at the present time. Under this power, if they had the right to refuse registration, we would not be able to sell in a lot of those States our materials. It is a very dangerous precedent."[6] Again, Hitchner used state sovereignty to set independent directions for use (and the ability of companies to market insecticides accordingly) to undercut registration and centralized authority.

Representative Walter K. Granger of Utah turned this argument on its head, when he commented on Hitchner's statement: "It seems to me there is another horn to that dilemma, too. I assume that the Secretary Agriculture, before he would disapprove the registration, would have competent chemists—I assume he would—to ascertain what it was, and

118

WHAT'S THE RISK?

instead of taking the bureau's idea and their chemists, the public would be forced to take what your chemists said; would they not? That would be the case of another individual saying what the situation should be."[7] Hitchner continued to resist central authority: "There you have an awfully concentrated amount of authority in one man."[8] But not all organizations were so resistant to further regulation of insecticides. Russell Smith of the National Farmers Union urged the committee to report the bill without substantial amendment, and he praised the extension of the bill to cover rodenticides and for the secretary of agriculture to provide a definition of what constituted "pests." Finally, he appreciated additional labeling safeguards, particularly the designation "highly toxic to man."[9]

Several speakers attributed the need for new legislation to the increasingly "scientific" nature of new insecticides. Dr. E. L. Griffin, who was assistant chief of the Insecticide Division, Livestock Branch at the USDA, referred to technical (or scientific) challenges in voicing his support for the new act: "The insecticide, fungicide, and rodenticide business has changed very markedly since 1910. It is now a highly scientific business. The products coming on the market are new and unknown products in a good many cases, and in our opinion they need a lot more careful supervision than is possible under the present act. We believe that this act should be brought up to date, and we believe that this is a good act."[10]

When the new insecticide bill (H.R. 1237) came up for debate in the House in May 1947, August Andresen (Republican, Michigan) introduced the bill and noted that it had the support of the insecticide industry, distributors, USDA, and farmer's organizations. Some of the manufacturers were resistant to registration, but for the protection of the public, this was a necessary part of H.R. 1237.[11] The only interchange of note occurred when Representative Frank B. Keefe (Wisconsin) asked why administration of the act would fall to the USDA rather than the FDA, which already administered the FFDCA. Andresen deflected Keefe's question by pointing out that the USDA administered the Insecticide Act of 1910. Keefe pressed his point noting that the new act would require separate testing facilities in two agencies. The chairman of the Agriculture Committee, John W. Flannagan, Jr. (Virginia), suggested that the act would primarily affect farmers and that it was currently administered by the USDA in the form of the Insecticide Act; the new act

119

only amended the old one. Without further debate, H.R. 1237 passed the House and went on to the Senate, where it passed without further debate.[12]

The Federal Insecticide, Fungicide, and Rodenticide Act, signed into law in 1947, dictated the licensing of the so-called economic poisons prior to their sale in interstate or international commerce. The law also stipulated that prominent warning labels detailing instructions for use be included on highly toxic pesticides. Furthermore, FIFRA required manufacturers to color powdered insecticides to prevent confusion with other household products, for example, flour, sugar, baking soda, and salt. Adelynne Whitaker emphasized that FIFRA assured consumers of quality pesticides while protecting them from accidental poisonings. Registration required manufacturers to test insecticides to determine efficacy before marketing their products. Yet public health officials were disappointed that FIFRA's registration clause did not reinforce the FFDCA and control pesticide residues. In his critique of an earlier version of the bill, Paul A. Neal of the PHS recommended a consideration of public health aspects and called for coordinating toxicity testing between PHS, FDA, and USDA.[13] FIFRA in its final form did not incorporate Neal's recommendations.

Though federal officials, consumers, and even manufacturers questioned the efficacy of FIFRA in addressing potential damage to the environment and health, like the Federal Food, Drug, and Cosmetic Act of 1938, FIFRA served as a critical initial step in the development of more comprehensive regulation. Nevertheless, according to environmental scientist John Wargo, the primary risk-management strategy reinforced by FIFRA after World War II was simply labeling. By requiring labels with instructions for use, the law implied that those who used pesticides could avoid adverse effects by following the directions. Despite the clear notes of concern regarding DDT and other new pesticides sounded during the FIFRA hearings, the law failed to address potential risks to health and the environment.[14] Wargo criticized the legislation as preferential to pesticides manufacturers: "This approach may have done far more to protect the entitlements of the pesticide manufacturers rather than either public health or environmental quality. It sheltered manufacturers from uncoordinated state regulations, and may simply have served to provide the public with a false sense of security that pes-

WHAT'S THE RISK?

ticide risks were being well contained by USDA. The reality was that USDA registered pesticides whenever asked."[15]

By 1951, concern regarding possible risks associated with insecticides and other chemicals that were finding their way into the food supply inspired a new round of congressional hearings before the House Select Committee to Investigate the Use of Chemicals in Food Products, which took place in the nation's capital and around the country from April 1951 to March 1952. Congressman James J. Delaney of New York chaired the hearings, which became known as the "Chemicals in Food Products" or the "Delaney Hearings." Another key member of the committee was Nebraska Congressman A. I. Miller. Yet it was the committee's chief counsel, Vincent A. Kleinfeld, who examined witnesses utilizing his comprehensive and encyclopedic knowledge of the FFDCA and its amendments as well as the prepared statements and exhibits.

Over the course of many days of hearings in several venues, including Washington, DC, Washington state, and California, with a transcript of more than 2,700 pages, several critical issues emerged. Congressmen repeatedly returned to their concerns regarding the widespread use of DDT and its potential health effects. Nevertheless, scientific uncertainty permeated the hearings. The views of scientists and public health officials on DDT and other synthetic insecticides ranged widely. Some scientists noted the lack of concrete evidence of harmful effects associated with DDT, and others cited anecdotal evidence of effects from mild to profound in connection with DDT and other chemical insecticides. Yet most participants agreed that the new insecticides had significantly boosted crop yields and contributed to public health since their intro duction in the years following World War II. Thus the Delaney Hearings provide a useful index to congressional interest in pesticides, broad-based uncertainty among scientists, and a sense that such chemicals had quickly emerged as critical to American food production and public health. There was, however, general recognition, particularly among state health officials, that the Insecticide Act of 1910 and the FFDCA of 1938 required revision.

In his analysis of the significance of the Delaney Hearings in the development of pesticides policy, Christopher Bosso has argued that industry initially resisted the hearings on the deeply entrenched view that any publicity constituted negative publicity. In the case of the Delaney

121

WHAT'S THE RISK?

Hearings, Bosso recognized that industry fears were justified, given Delaney's stated intent for the committee: assess possible dangers to public health and decide whether the threat warranted federal regulatory reform.[16] In revisiting the Delaney Hearings, I point to how scientists and those charged with the responsibility to evaluate scientific evidence evaluated risks associated with exposure to pesticides, particularly chlorinated hydrocarbons and organic phosphates, which had proliferated since the passage of FIFRA in 1947. To a certain degree, the views of witnesses correlated to their primary association. To the surprise of no one, industry and USDA representatives defended the value and safety of new insecticides, but there were notable exceptions: scientists from academia, the PHS and the NIH conveyed considerable uncertainty regarding risks in the face of the committee's pointed questioning.

One of the first witnesses proclaimed the importance of new insecticides to the food supply. K. T. Hutchinson, who was assistant secretary of agriculture, quantified losses due to pests: "Notwithstanding the large losses caused by pests, which for insects alone are estimated to approximate $4,000,000,000 annually, without the use of pesticides it would be impossible to produce needed food. The importance of controlling pests is being recognized, and more farmers and livestock producers are using the essential aids to protect the investments they make to produce crops and domestic animals. That the control of pests is one of the most effective ways to increase yields and supply high-quality products is becoming generally accepted."[17]

Numerous state health officials submitted statements to the hearing. Most focused their concern on additives and substitutes that were being introduced to food products deliberately, but some also cited the potential risks of chemicals that entered the food supply inadvertently, such as insecticides. R. L. Cleere, executive director of the Colorado Department of Health, articulated worries about the pesticides: "Our department is in accord with any legislation to control the use of harmful chemicals which may find their way to the consuming public in foods. Today many new insecticides, rodenticides, and fungicides are being used which are definitely toxic and their toxicity levels on the end products have not been ascertained."[18] In addition to uncertainty regarding the new insecticides, Cleere noted the lack of legislation in Colorado:

122

WHAT'S THE RISK?

"Colorado has no definite legislation as yet on this matter. There has been no organized work of a State nature on toxicities. Our food and drug laws do prohibit poisonous, adulterated foods, knowingly marketed as contaminated by poisonous or harmful substances and resulting in sickness or death from eating such articles of food or drink. This would apply to known poisons; but what about some of our present day insecticides, etc.?"[19] Another state health official, Carl E. Weigele of New Jersey, recommended extensive testing before market release.[20] Interestingly, Weigele used the New Drug section of the FFDCA (§ 201 (p)) to support the practicability of such toxicity testing without burdening private industry. George A. Spendlove of the Utah Department of Health expressed similar sentiments in this statement: "Our department feels very strongly that Federal legislation should be enacted providing, as in the case of new drugs, that chemicals introduced in foods shipped in interstate commerce should first be demonstrated to be safe to the satisfaction of the FDA."[21]

One of the clearest (and most alarming) presentations on DDT came from a Connecticut physician named Morton S. Biskind. Before receiving his M.D. at Case Western Reserve University in 1930, Biskind earned a master's degree in pharmacology in 1928. Though he served as a research fellow at Case Western Reserve University, as a member of the headquarters staff of the council on pharmacy and chemistry in the AMA, and as head of the endocrine laboratory and the endocrine clinic at Beth Israel Hospital in New York, Biskind restricted his statement to his direct clinical experiences with patients over the course of two and a half years. He argued:

The introduction for uncontrolled general use by the public of the insecticide DDT, or chlorophenothane, and the series of even more deadly substances that followed has no previous counterpart in history. Beyond question, no other substance known to man was ever before developed so rapidly and spread indiscriminately over so large a portion of the earth in so short a time. This is the more surprising as, at the time DDT was released for public use, a large amount of data was already available in the medical literature showing that this agent was extremely toxic for many different species of animals, that it was cumulatively stored in the body fat

WHAT'S THE RISK?

and that it appeared in the milk. At this time a few cases of DDT poisoning in human beings had also been reported. These observations were almost completely ignored or misinterpreted.[22]

Biskind proceeded to describe what he called "Virus X," a syndrome comprising many symptoms—acute gastroenteritis, nausea, vomiting, abdominal pain, diarrhea, running nose, cough, sore throat, pain in the joints, muscle weakness, insomnia, headache, hypersensitivity, numbness, twitching of voluntary muscles, loss of weight, and psychological effects, among others.[23] Several of Biskind's patients had complained of these symptoms in association with encounters with DDT and other pesticides. Biskind had described these cases in a series of articles published in 1949.[24]

In the midst of Biskind's statement, Congressman A. I. Miller, who was also a doctor, interrupted to ask if Biskind had published these findings in scientific medical journals. Biskind responded that he had, but Miller countered by citing the findings of the AMA: "Generally the theory that DDT is highly toxic in concentrated doses is accepted, of course. But the American Medical Association, a group that goes through these poisons with a fine-toothed comb, has not reached the conclusions that you are now giving to the committee. Is that true?"[25] When Biskind replied that he was aware of the findings, Miller asked if he could identify any large group of scientific men who accepted Biskind's conclusions. After admitting that he could not, Biskind cited three scientists who agreed with him. Miller then queried the chairman whether the record should be encumbered with something of doubtful standing as far as scientific men were concerned. Delaney countered that Biskind should read his statement. Miller noted his reservations once again: "It is true that what he says has not been accepted by a majority of the scientific men. I maintain that there is only a very small segment that accepts this viewpoint; and if there were a large segment that accepted the viewpoint, then the Government would have no right to permit DDT to be used any place. It is their fault, if they accept this man's viewpoint and findings. I am inclined to be sympathetic with him, because I think there is something to it, but if what he says it true, then it goes counter to the other large group of scientific men that says that it is safe to use, and has been given the green light by Government agencies."[26] Before

124

WHAT'S THE RISK?

Biskind could resume, Congressman E. M. Hedrick, another physician, questioned Biskind: "As a matter of fact, lots of people have been exposed to DDT with apparently no injurious effects; is that not correct?"[27] Biskind answered it was true. Hedrick wondered whether the people Biskind had described had been hypersensitive to DDT. Biskind replied: "I do not think that there is any question that they are. But I think that the number that are, is far larger than ordinarily supposed."[28]

Biskind resumed his statement by describing several additional patients who had experienced a variety of symptoms in association with exposure to DDT through various pathways. Patients had encountered DDT in their foods, as dusts in aerosol sprays, and even from wallpaper and clothing impregnated with the chemical. Biskind based his original research on more than two hundred cases, but he had learned of many additional cases. He argued that exposure to DDT was virtually universal and that it was impossible to separate the effects of direct exposure and those that occurred following ingestion of contaminated food. Even specimens of mother's milk from patients with a history of exposure showed DDT. Cow's milk offered no alternative, Biskind argued, since USDA reports indicated that samples contained 0.5 to 25 ppm of DDT.

Biskind worried about other chemicals, including chlordane, BHC, and parathion. He acknowledged that these chemicals posed a dilemma: "We are dealing with double-edged swords, for the very substances now promoted to increase the size of our crops in the long run turn out to be detrimental to agriculture itself. All these substances and the fantastically toxic parathion, too, inhibit the growth of certain plants, and compounds of the DDT group also persistently poison the soil, so far as present evidence goes, for 5 or 6 years and possibly indefinitely."[29] The dire threat of parathion, in particular, extended beyond the risk to humans: "Parathion is everywhere admitted to be deadly for man and all other animals. One manufacturer warns that sprayed areas may not even be entered with out a mask and protective clothing for 30 days after application. Failure to heed this precaution has already resulted in numerous serious accidents to men. What happens to the birds and other wildlife who cannot read?"[30]

In his conclusion, Biskind exhorted Congress to take action against the use of certain pesticides on crops: "It is my opinion that the use on crops or in food establishments of any sort, of the chlorinated cyclic

125

WHAT'S THE RISK?

hydrocarbons—which include the DDT group of compounds and the organic phosphates of the parathion group—should be, and, if we want to survive, must be—specifically forbidden by law."[31] In response, the chief counsel, Kleinfeld, asked a few questions beginning with, "Doctor, I think you testified that the views, which you have here expressed are not generally recognized by the medical or scientific profession; is that correct?"[32] Biskind admitted that this was true. Kleinfeld then asked Biskind to cite his publications regarding DDT and to indicate any other scientists who had conducted experimental work on the insecticides he discussed. Finally, Kleinfeld requested that Biskind read aloud the text of a press release issued on April 1, 1949, by the Federal Security Agency and the USDA after a meeting that included representatives from appropriate divisions of the U.S. Army and Navy as well as the FDA, the PHS, the Office of the Surgeon General, and the Pan American Sanitary Bureau. The statement dismissed concerns regarding DDT: "It is well recognized that DDT, like other insecticides, is a poison. This fact has been given full consideration in making recommendations for its use. There is no evidence that the use of DDT in accordance with the recommendations of the various Federal agencies has ever caused human sickness due to the DDT itself. This is despite the fact that thousands of tons have been used annually for the past 4 or 5 years in the home and for crop and animal protection."[33] After a few more questions of a general nature, the committee dismissed Biskind.

For information regarding the toxicity of DDT to humans, the Delaney Committee called on Wayland Hayes and Paul Neal of PHS. In their statement, Hayes and Neal acknowledged that DDT was a highly toxic in large amounts and noted the symptoms of acute toxicity.[34] Having clearly established the acute toxicity of DDT to animals, Hayes reviewed the medical literature regarding its toxicity to humans, including his inhalation and ingestion experiments (see chapter 2). From this research, Hayes noted: "No objective or subjective symptoms were found in spite of thorough physical examinations, including neurological, biochemical, hematological, psychophysiological, electroencephalographic and electrocardiographic studies."[35]

Despite the extensive testing with animals, Hayes and Neal stated, there was no evidence of harmful effects of the ingestion of DDT: "We

126

have found that although a great deal of animal experimentation has been carried out with DDT there are no bona fide scientific reports of human cases following the ingestion of small amounts of DDT, although, as just noted, acute poisoning following large doses has been encountered."[36] With this statement and many others, the PHS scientists directly challenged Biskind's testimony. To contradict Biskind, Hayes and Neal deployed several strategies. First, they undermined the scientific validity of two British studies on which Biskind based his symptomology for DDT by questioning the methodology, the reproducibility, and the general environment (postwar England). But Biskind's anecdotal evidence of DDT poisoning in his patients struck Hayes and Neal as particularly problematic. For example, if more than a third of Biskind's patients exhibited symptoms associated with DDT poisoning, as he testified, Hayes and Neal rhetorically wondered why other doctors were not reporting similar numbers of poisonings in their patients. To counter Biskind's claims, they reported on a surveillance program that began in 1945, when DDT was released for public use, and continued until November 1947. Of forty reported cases, none yielded a diagnosis of DDT poisoning after toxicological investigation and hospitalization, in some cases. Moreover, they reported, "To the best of our knowledge, there have been no substantial cases of DDT poisoning in this country resulting from the ingestion of food containing DDT as a residue."[37] While acknowledging DDT's significant toxicity, Hayes and Neal reiterated and expanded this point in their conclusion: "In summary, it must be emphasized that DDT is a toxic substance and can cause injury if not properly handled. Undisputed cases of acute illness have been reported. It is also true that there are accurate reports of the presence of DDT in the body fat and milk of human beings. There is, however, no authentic report of liver injury or other chronic poisoning in man resulting from DDT."[38]

Hayes and Neal, who fielded many of the questions over the course of the hearing, yielded little ground from their initial assertion of the safety of DDT as a residue in food in low quantities. Even in the case of the USDA's recommendation in 1949 to limit use of DDT in dairy barns, Neal pointed to the lag between experimental work and publication as one reason the PHS did not have information on the storage of DDT in fats. But Kleinfeld pressed the point:

WHAT'S THE RISK?

KLEINFELD: "And I suppose that is why no objection was taken to the use of DDT in dairy barns and on dairy cattle; is that correct?

NEAL: "I think that is a question for the Department of Agriculture, but from my memory I can tell you, sir, what I remember of it. It was that if you used DDT in the barns in dairies and you kept it off the cow's food it was thought that very little absorption would occur."[39]

There were other instances wherein Neal or Hayes directed Kleinfeld to another branch of government or dismissed a question as falling outside the realm of science. For example, Kleinfeld questioned Hayes about a recent paper in the *Journal of the American Medical Association* that suggested the possibility that DDT and other chlorinated hydrocarbons had adverse effects on the functions of adipose tissue. Kleinfeld quoted from the paper and asked Hayes if he agreed with the statement. Hayes replied: "I agree with the statement, but you notice he says they *may* have."

KLEINFELD: "Yes. I asked if you would agree with that statement— that it may have.

DR. HAYES: Yes; it may have

MR. KLEINFELD: Pardon me?

DR. HAYES: I say it may have. But there is no proof in the literature that it does.

KLEINFELD: Is there any proof that shows that it does not?

HAYES: No. It is just that we are quoting from a scientific journal and they are discussing the possibilities.

KLEINFELD: That is correct.

HAYES: And that is one of the possibilities which must be considered.

KLEINFELD: Which should be considered before a chemical is used on a food product?

HAYES: I believe that is a legal question, not a medical one.

KLEINFELD: I would think the ordinary person would answer that question, sir; but if you do not want to, let me ask you another question.[40]

128

Having raised scientific uncertainty to qualify the findings of a paper that could have important implications for the food supply, Hayes dismissed the important questions of what should be considered before a chemical is used on a food product as a legal question (and implied that such questions fell outside his purview).

At other times, Hayes deflected questions by referring to existing federal legislation or recommendations. After again referencing the *J.A.M.A.* article, Kleinfeld asked Hayes if he believed that DDT should be used in dairy barns. Hayes responded, "That has already been prohibited after a decision made jointly by the Food and Drug Administration, Department of Agriculture, the Public Health Service, and perhaps others."[41] Similarly, Hayes deflected a question regarding the development of resistance to DDT in houseflies, noting that he was no entomologist, after he acknowledged that researchers in the U.S. and abroad had revealed that houseflies did develop DDT resistance.

Wayland Hayes's testimony before the Delaney Hearings was vitally important in part due to his position as the chief of toxicology for the PHS. Few government scientists had better access to the full range of data regarding the potential health effects of DDT and other chemical insecticides. No one was better placed to provide Congress with a clear sense of potential human health effects of exposure to DDT. Hayes made it clear that DDT was a highly toxic chemical that in large doses could sicken humans. He acknowledged acute effects, but he knew of no definitive studies that had shown chronic effects in humans as a result of small exposures. Although he conceded that the USDA, FDA, and the PHS had recommended against the use of DDT in dairy barns, Hayes offered no indication of his view of this action though surely concern regarding chronic effects of exposure to DDT in cow's milk inspired the recommendation. By parsing questions as scientific or legal, concerned with other scientific specialties, or covered by existing legislation, Hayes characterized (in one individual) exactly the fissures that delayed a coordinated response to the risks of chemical insecticides, let alone legislation. As the hearings proceeded, other scientists would present potential DDT residues as a real threat to public health leaving Congress no clear path toward reasonable legislation.

One of the recurring issues as the hearings proceeded was the considerable gulf that existed between science and policy. Scientists accepted a

WHAT'S THE RISK?

degree of uncertainty, while legislators sought "proof" of threats to human health and safety before enacting laws restricting use. Occasionally members of the committee expressed their frustration over the failure of scientists to clarify the safety of DDT. As part of his prepared statement, Frank Princi, who was an associate professor of industrial medicine at the University of Cincinnati, endeavored to capture for the committee the scientific uncertainty related to DDT: "[DDT] has been subjected to more scientific investigation than any other organic material. Yet, despite this knowledge, there is still sharp disagreement concerning the hazard associated with its use. On the one hand, we are told that it is the safest of insecticides; and, on the other, it is suggested that its toxicity may have been underestimated and that it is probably responsible for such conditions as suicidal tendencies, aplastic anemia, pneumonia, leukemia, virus X, arteriosclerosis, and even cancer."[42] Princi's comment reveals the wide range of conditions anecdotally associated with DDT, but he identified extrapolation from animal data to humans and real-life exposures as a critical element of scientific uncertainty: "Much of this controversy had developed because of attempts to translate the results of animal experimentation into human experience without appropriate consideration of the variability of animal species. *Other diversities of opinion have developed because of a lack of understanding of the actual conditions of exposure which result from ordinary methods of use of the material.* It is suggested that these questions cannot be resolved fairly and adequately by any single government agency."[43] Insofar as gaps existed between different scientific specialties, there were also fissures between government agencies. Regulation of problems that cut across different agencies proved particularly difficult.

It was however, Princi's comment about the scientific uncertainty surrounding DDT that drew the ire of A. I. Miller. Miller initially requested from Princi a simple clarification: "You state, of course correctly, that there is sharp disagreement among the experts," to which Princi responded, "Yes, sir." At that point, Miller's frustration boiled over and he responded sharply: "That does not help this committee. We are at sea when experts disagree. Why cannot we get some experts up here who can say DDT is or is not harmful under certain limitations? Have you any opinion, as an expert, on the question of DDT?"[44] Princi replied, "In my opinion DDT, in the manner in which it is now used and

130

WHAT'S THE RISK?

in the quantities to which persons are presently exposed, is apparently innocuous."[45]

Miller again returned to the problem of scientific uncertainty, when he asked if Congress would be on sound ground if it said to insecticide manufacturers that "[they] must prove to the satisfaction of either a group of scientists that might be set up that the chemicals are not harmful to the human being before they are to be used in commercial food for the public." Princi took exception to the word "proof": "It is difficult for me to answer that question since you have used the word 'proof' because . . ."[46] Miller interrupted Princi: "Then let's put it another way. Who should have the burden of proof that it is not harmful?" With the question framed in a way to avoid the issue of scientific uncertainty, Princi replied decisively with a concise iteration of the precautionary principle: "The burden of proof certainly should fall upon the supplier or manufacturer. I think there is no question on that."[47]

Scientific uncertainty also proved challenging beyond the realm of human health effects. The problem of insects developing resistance to DDT arose in the testimony of Charles E. Palm, who was a professor of entomology at the New York State College of Agriculture at Cornell University. In his prepared statement, Palm underscored this problem, noting that DDT had become ineffectual against houseflies in New York dairy barns as a result of evolving resistance. He worried that houseflies were developing resistance to other insecticides too.[48] When pressed, Palm noted that he had found flies resistant to the recommended dosage and even much higher dosages. Moreover, the Cornell entomologist noted that resistance to one chlorinated hydrocarbon conferred a measure of resistance to all chemicals in the class.[49]

Several of the members of the committee seemed to fully appreciate the implications of Palm's statement; namely, that flies could develop resistance to all of the insecticides within the class of chlorinated hydrocarbons. The phenomenon was not restricted to New York but was a general difficulty across the United States. Palm indicated that resistance had developed in as little as three years.[50] Congressman Walter Horan noted that such was the case in his home state of Washington.[51] In response to further questioning regarding insecticide resistance, Palm introduced the committee to two concepts—the "balance of nature"

WHAT'S THE RISK?

and biological control—only to dismiss them in favor of DDT, which growers found more reliable.[52]

Like other witnesses, Palm argued that agricultural productivity in the United States depended on chemical insecticides. He also believed that insecticides were "chemical protectants used in the production and protection of food and not as chemical additives." Entomologists could make recommendations concerning the type, quantity, and timing of particular insecticides to minimize residues and risks to consumers. As one example, Palm cited increased yields of Irish potatoes without evidence of DDT residues. He also noted that entomologists consulted with the FDA and the USDA regarding problems, but he stressed the difference between insecticides and drugs (also regulated by the FFDCA of 1938). Palm exhorted the committee to maintain existing distinctions between insecticides and drugs in order to provide farmers with insecticides quickly. Public health could be protected with the addition of research facilities, which would decrease the time to appraise health hazards while insuring "the use of pesticides in their beneficial roles of providing adequate supplies of food and fiber as well as essential roles in reducing insect vectors of pathogens causing diseases of man and animals."[53] With his final statement, Palm spoke for many, if not all, economic entomologists. Namely, chemical pesticides provided extraordinary benefits in the service of both public health and food production. Risks could be managed with additional research facilities at least in the theoretical sense.

Another economic entomologist, George C. Decker, who was head of the Economic Entomology Section at the Illinois Natural History Survey and the Illinois Agricultural Experiment Station, presented a view largely in accordance with Palm's. Decker, however, underscored the general lack of evidence regarding health risks associated with insecticides. He noted that despite the use of millions of pounds of pesticides each year for the previous five years, there were very few recorded deaths attributable to insecticides and all were due to operations hazards or misuse. The general counsel, Kleinfeld, pressed Decker to acknowledge potential criteria of safety other than death, and he asked whether it was possible to say that no acute or chronic illnesses had been caused by the annual use of millions of pounds of insecticides. Decker answered the question in the abstract: "No one can answer that question with assur-

132

WHAT'S THE RISK?

ance and certainty. The circumstantial evidence would indicate that there is little, if any, of that, in my opinion. It seems to me, sir, that in studying accidents of any kind, the fatalities are an index of the other injuries."[54] Kleinfeld persisted in pressing the point that there could be illnesses in the absence of fatalities and that these illnesses might not register in vital statistics or even newspapers for a long time. Decker held his ground, arguing that any serious problems should have appeared in the media in the four to five years since the introduction of DDT.[55] Still, Kleinfeld argued the subtler point regarding the possibility of chronic illnesses associated with the ingestion of very small amounts of an insecticide over a very long time, to which Decker finally acceded: "That is correct. I cannot deny that."[56]

Up to this point, one can certainly understand why members of the Delaney Committee found themselves in a state of confusion regarding the risks of chemicals, particularly insecticides, in food products. In point of fact, PHS toxicologists and economic entomologists demonstrated a considerable degree of agreement regarding the significant benefits and the relatively minor and, for the most part, manageable risks associated with insecticides. Thus the testimony of John Dendy, head of the Analytical Chemistry Division of the Texas Research Foundation in Renner, Texas, may have caught the committee off guard. Dendy concluded his brief prepared statement with four conclusions: there was widespread contamination of both animal and human foodstuffs with DDT and other chlorinated hydrocarbons, contamination was spread and intensified by the continued use of chemical insecticides, continued use of DDT and other chlorinated hydrocarbons posed an ever-increasing hazard to the public health, and existing laws and/or enforcement procedure were insufficient to prevent the development of this serious hazard to human health. He asserted that the Texas Research Foundation planned to continue its research into these conditions.[57] E. H. Hedrick of West Virginia, expressed his appreciation for Dendy's clear exposition: "For the record, that is about as forthright a statement on DDT, about the results of DDT on human beings as I have ever heard."[58] And yet, when Hedrick asked Dendy if the present law was sufficient, Dendy noted that he was not a lawyer (and by intimation unqualified to render an expert opinion on legislation), but he could comment on the law's inadequacy in its original writing for allowing

133

WHAT'S THE RISK?

indiscriminate use of insecticides or in its enforcement for permitting contaminated products to be consumed by individuals.[59]

Just as he hesitated to give his opinion on legal matters, Dendy also deferred to doctors on matters concerning human toxicity. When Congressman Miller asked him what concentrations of DDT would be injurious to humans, Dendy noted his lack of medical credentials, but he then proceeded to elaborate on the implications of biomagnification: "Well, that puts me on a spot, not being a medical man, and to date no one yet has established the so-called LD-50 evidence in all these insecticides, because, first of all, in any specific species, whether they are rabbits, white mice, or human beings, each individual has a specific tolerance and it does not exactly correspond to its next-door neighbor, even its litter mate, so an LD-50 is difficult to establish, and men who are well qualified have not established that. Milk containing small concentrations of DDT has been found by most of the investigators in the field. Even though the intake is small, the fatty accumulation in the tissues as the result is magnified as high as 34 times the original intake. In other words, with a diet of 10 parts per million you could expect, in some instances 340 ppm."[60] By suggesting that LD_{50}s had not been established for all insecticides and that they could vary across species and even individuals, Dendy turned the discussion back to scientific uncertainty, but he reframed the problem in terms of its implications for human health by explaining how DDT magnified within organisms, including humans, which meant that small exposures (10 ppm) could build up in tissues to levels as high as 340 ppm. Dendy's testimony provides a very early example of a specific concern—biomagnification—regarding the environmental implications of the widespread use of DDT and other insecticides.[61] In these comments, he anticipated one of Rachel Carson's central arguments in *Silent Spring*.

Moreover, in another prescient statement, Dendy noted the gaps between professions, for example, between chemists and physicians. Specifically he wondered when the medical profession would indicate whether or not DDT produced death in human beings. Despite extensive research conducted on the detection of DDT, as a chemist, Dendy believed that he had struck a barrier in the medical profession. In Dendy's opinion, lack of coordination contributed to the problem. He elaborated on this point: "Yes, remember, sir, there has been much work done on it,

134

but the lack of coordination of the individuals doing the work, their unwillingness to share information with one another, has been the chief draw-back. There are only two sides of this fence. You have to talk relatively freely with those individuals who are on your side of the fence and those who are on the other side of the fence are rather hesitant, and this was our objective."[62]

The benefit of insecticides to agricultural productivity was a consistent undercurrent during the hearings, but the Delaney Committee also heard from at least one farmer who wondered about the risks of the new insecticides. In his prepared statement, Louis Bromfield, the owner of Malabar Farms in Lucas, Ohio, argued that the effects of new insecticides on humans and animals were largely unexplored: "Certainly their use should raise grave doubts. Put in the simplest terms, what is poisonous to the organic structure of an insect must also be poisonous in sufficient immediate quantities or in sufficient accumulated quantities to other life as well."[63] Bromfield could testify to the benefits of insecticides, but he found the risks to be more worrisome, arguing that the nation had "plunged into the wholesale use of all these poisons with little or no research concerning their ultimate effects upon health, vitality, and the powers of reproduction" to the detriment of "virtually every citizen."[64]

Unlike other farmers who testified, Bromfield spoke as someone who had used insecticides and as a citizen concerned about their rapid proliferation and potential health effects. He implicated the chemical manufacturers for their blind promotion of insecticides. Meanwhile, he was aware that insecticides could lose their effectiveness as insects developed resistance (or immunity). When Kleinfeld asked Bromfield whether an overuse of insecticides upset the natural balance of nature, Bromfield relayed the tendency of insecticides to kill all insects, harmful and beneficial alike. He noted that destroying ladybugs led to an explosion of aphids. Furthermore, it appeared that many birds and fish were dying as a result of spraying. As birds and beneficial insects died off, the insect population could double, triple, and even quadruple.

Despite his evident reservations regarding the overuse of DDT, Bromfield resisted its removal from the market or restrictions on its sale. Limited use of DDT in specific areas (feeding barns and loafing sheds) could be effective and safe, according to Bromfield. He extended the

WHAT'S THE RISK?

notion of specificity to target insects also. DDT, Bromfield acknowledged, was or had been in the past a "great fly killer," but he had de-
cided to shift to chlordane since DDT had become almost ineffective.
Bromfield's testimony clearly reflects his deep-seated ambivalence regarding DDT and other synthetic insecticides. The rapid development
of resistance in flies and other target organisms seemed to necessitate
application of greater concentrations with increasing frequency, both of
which meant greater exposures. Bromfield attempted to minimize these
risks by limiting his application of DDT to specific areas and by target-
ing specific pests (predominantly flies and lice).[65]

Members of the Delaney Committee must have found the wide dis-
parity in testimony regarding DDT confusing. Many of the witnesses
deferred to physicians and toxicologists to clarify what risks existed and
their severity. Yet when individuals who studied human health effects
testified, they couched their statements carefully. For example, Harold
P. Morris of the nutrition unit at the National Cancer Institute at NIH
noted the challenges to establish whether a particular compound induced
cancer in animals. Morris concluded: "In summary, I have pointed out:
(1) That a large number of chemical compounds induce cancer in animals. (2) That there is no way of predicting their cancer-inducing properties without a biological test. (3) That the careful testing of chemicals
for cancer-producing properties in animals is exceedingly difficult to
evaluate. Any test for cancer is influenced by a very large number of environmental and hereditary factors which the experimenter must seek to
control and evaluate."[66] Despite the difficulties inherent to the analysis
of compounds for carcinogenicity, Morris believed that any estimate of
the possible injurious properties of chemicals added to nutrients should
include testing for cancer-causing properties in several species of animals
prior to approving their use in food. With that said, he sharply criticized
a recent article in the *British Medical Bulletin* regarding the carcinogenic
action of heated fats and lipoids on the grounds that the researchers used
the rat as their model. Due to the nature of its stomach, the laboratory
rat was unsatisfactory, and Morris discredited studies that used rats. As a
result, Morris's lab had never successfully reproduced the results of the
British study. Such statements must have added to the committee's growing sense of confusion in light of the fact that the vast majority of laboratory toxicity studies utilized rats, mice, rabbits, or dogs.

136

The committee heard from another expert on environmental cancer: Wilhelm C. Hueper, chief of the Cancerigenic Studies Section of the Cancer Control Branch of the National Cancer Institute and chief of the associated laboratory. More important, Hueper was widely regarded as a leading expert on occupational and environmental cancers (see chapter 2). But when asked in what capacity he was appearing before the committee, Hueper replied that he was testifying as a private citizen, basing his testimony on his experiences and training of the previous twenty-five years. In response to Miller's questioning, Hueper noted that cancer incidence was rising in general due to the growing population and the increase of older age-groups in the population, but both lung cancer and leukemia had increased in recent years. Men seemed to be particularly vulnerable to lung cancer, presumably as a result of occupational exposures. Hueper proceeded to review specific dyes and other chemicals, including arsenic, that caused cancers. He noted that arsenical insecticides could be carcinogenic if levels rose to chronic arsenicism. Turning to chlorinated hydrocarbons, Hueper adopted a cautious tone. Kleinfeld asked, "What is the evidence, if any, that these chlorinated hydrocarbons may be carcinogenic?" Hueper replied: "I think we have to get away from the word 'carcinogenic' here. We have to use a more neutral term and say 'tumor-producing,' leaving it open whether the tumors produced are actually carcinomas or not, or cancers or not."[67] When Kleinfeld asked him to identify the lowest level of chlorinated hydrocarbons to cause tumors if ingested, Hueper stated: "I think I should emphasize that we have no record of human cancer from exposure to DDT, although we have evidence of cases of severe liver poisoning from exposure to other more powerful liver toxic agents like chloroform and carbon tetrachloride, but none of those chlorinated hydrocarbons so far, as far as we know, has caused cancer in men."[68]

Despite Hueper's refusal to directly link chlorinated hydrocarbons to cancer, Kleinfeld continued to press, citing considerable evidence that DDT was accumulating in the fat of people not directly exposed to it. Kleinfeld asked: "Would such an accumulation of DDT in the fat of the ordinary person, young or old, well or sick, possibly create some hazards?" Hueper replied: "I think on the basis of the evidence we have right now we cannot say. We have to wait, perhaps, *10 or 15 years* to see whether such evidence may be forthcoming."[69] It is fair to say that no

WHAT'S THE RISK?

one in the U.S. was better positioned than Hueper to render an informed opinion on the possibility that chlorinated hydrocarbons might cause cancer, but as of January 1952, he believed that such a case would require ten to fifteen years of research. Historian Robert Proctor has suggested that the medical profession regarded Hueper as a maverick in his unrelenting effort to track down industrial carcinogens. More significant, Proctor noted that Hueper's views had lost favor in the realms of science and politics. By the 1950s and 1960s, medical researchers regarded environmental carcinogenesis as rather out-of-date; researchers considered nonchemical factors, such as viruses and genetics, as more significant in the etiology of cancer. More important, according to Proctor, in the era of postwar conservatism, Hueper's prolabor and perceived anti-industry stance of cleaning up the workplace and the environment garnered little support.[70]

In 1948, Hueper became the founding director of the Environmental Cancer Section of the National Cancer Institute (NCI), the research arm of the PHS. Research Hueper had initiated before he achieved his position at the NCI resulted in his departure. Specifically, he accepted a consultancy at the Baltimore plant of the Mutual Chemical Company to investigate the link between chromium dust and lung cancer. The NCI funded Thomas Mancuso of the Ohio Health Department to study chromium dust at an Ohio plant. In 1951, Mancuso and Hueper jointly published a paper that confirmed elevated lung cancer rates at the chromium plant. When they attempted to publish another paper that suggested the possibility that the risk of cancer extended to the population outside chromium plants, the chief of the Industrial Hygiene Division at NCI ordered Hueper to remove his name from the paper. Hueper acquiesced but complained to the surgeon general, who shut down all of Hueper's activities outside the laboratory. Moreover, Hueper had to "discontinue work on chromium, end his field work, and cease all contact with industry and with state and local health agencies."[71] Environmental scientists Benjamin Ross and Steven Amter have suggested that the strictures placed on Hueper were the result of express or implied threats against other PHS programs (by the 1950s, Clarence Cannon, who in 1937 had redirected pesticide research from the FDA to the Industrial Hygiene Division, had become chairman of the House Appropriations Committee). When Hueper resumed research on chromium

138

WHAT'S THE RISK?

two years later, his superiors shut down his program after less than six months.[72] Frustrated by political pressures on his research and advocacy, Hueper resigned from his prominent position in cancer research in 1964.

Despite the professional challenges he faced throughout his career, Hueper provided critically important insights that would continue to resonate with legislative and regulatory efforts to control additives to foods.[73] When Miller asked him for his suggestions regarding the use of chemicals in foods, Hueper responded with a prescient statement that anticipated the direction of legislation and regulation: "I would feel that the uncontrolled use of any known or suspected agent with carcinogenic properties is not advisable, and that certain control measures should be taken."[74] On further examination, Hueper noted that the general population was probably exposed to materials in food more than through any other product, including cosmetics and medicines. On the basis of that view, he recommended toxicity testing for chemical additives to foods even in small doses.[75] Although the size of the committee was somewhat diminished on the day of Hueper's testimony, due to illness and other committee meetings, a few key policy makers took his recommendations to heart as they contemplated novel regulation.

Throughout the hearings, testimony kept returning to DDT and its potential toxicity for humans. Even the small sample of expert testimonies analyzed to this point reveals a wide range of views regarding risks associated with novel insecticides. Entomologists, toxicologists, and farmers presented widely divergent views with respect to DDT from fairly harmless, when used properly, to extremely toxic and becoming more so as the chemical accumulated within organisms. It has also become clear that the primary link between government regulators and end users was the USDA. Within the USDA, it was the Bureau of Entomology and Plant Quarantine, based in Orlando, Florida, that was responsible for the analysis of the new chemical insecticides. On May 22, 1951, the committee heard the testimony of Fred C. Bishopp, assistant chief of the BEPQ in charge of research. Two other scientists, Edward F. Knipling and W. C. Shaw, accompanied Bishopp to offer additional insights and clarifications. From the outset, Bishopp acknowledged concerns regarding the release of new and highly poisonous insecticides for public use, but like other entomologists, he suggested that they were

139

WHAT'S THE RISK?

comparatively safer than the insecticides they replaced: "Some people have been apprehensive of the release of new and highly poisonous insecticides for public use. Actually, many of these materials are no more poisonous than nicotine, arsenicals, and sodium fluoride that have been used as insecticides for many years. The newer materials—including DDT, benzene hexachloride (BHC), toxaphene, and chlordane—have, to a considerable extent, replaced these older insecticides and are used on a much larger scale. Nevertheless, probably fewer accidental deaths from acute poisoning by the new materials occur today than were caused by the older insecticides in the past."[76] Bishopp argued that new insecticides underwent greater scrutiny than in the past before a manufacturer would release the chemical to the public and before it would be registered by state or federal agencies. Despite field and laboratory studies that concerned formulations, mode of action, effectiveness under varying ecological conditions, toxicity to plants and animals, and spray residues, Bishopp acknowledged that such analysis did not necessarily cover all fields of public interest before the product became available for public use. Bishopp was confident that additional research would lead to the development of more efficient pesticides that were essential to achieve crop production requirements necessitated by the "national emergency."

When he turned to specific insecticides, Bishopp offered a clear statement of the potential risks of DDT: "Although certain information on new insecticides is lacking, one of the most pressing problems is the dissemination of the authentic available facts to the public. It must become more widely known that DDT and related compounds, although of relatively low acute toxicity to man, are persistent and therefore residues on crops must be reduced to a minimum."[77] With this statement, Bishopp addressed one of the constant sources of confusion regarding the toxicity of DDT; namely, that the persistence of the chemical within organisms had the potential to increase its toxicity over time. What Bishopp referred to as the "[organic] phosphate insecticides," like parathion and HETP, were quite different in that they were highly poisonous, but the residues rapidly degraded. It was the organic phosphates, he noted, that were responsible for most of the serious accidental insecticide poisonings that occurred in the previous two years, although he attributed them to workers failing to use respirators and protective clothing.[78] Thus, Bishopp characterized the trade-offs between DDT and other

WHAT'S THE RISK?

chlorinated hydrocarbons and the organophosphates. DDT presented relatively low toxicity, but it persisted and accumulated in the environment. Organic phosphates with high toxicity posed serious risks to operators, but given their rapid rate of decomposition posed minor risk to the food supply.

Like other experts, Bishopp testified to the extraordinary gains in food production that resulted from the widespread application of chemical insecticides, or rather, he suggested the abysmal state of crops in the absence of insecticides.[79] He also cited several specific cases in which DDT had significantly controlled an insect outbreak. A potential outbreak of velvetbean caterpillar in 1946 provided one example; Bishopp noted that prompt application of several insecticides lessened loses to several crops, but DDT produced faster action: "Dust mixtures containing from 2.5 to 5 percent DDT applied at rates of *12 to 20 pounds per acre* gave faster action against the caterpillars than cryolite or calcium arsenate and resulted in generally higher control."[80] In another case, insect control promoted tomato yields: "In California, during 1945, approximately 66,500 acres of tomatoes were treated by airplane with 10 percent DDT, at *65 pounds per acre,* in the dust form, for the control of the tomato-fruit worm, using 4,322,500 pounds of a DDT insecticide."[81] Certainly these two cases bolstered Bishopp's case that DDT provided effective control against insect outbreaks. And yet it is remarkable that Bishopp, the head of research at USDA's BEPQ, recounted spray campaigns that applied DDT at 12 to 20 pounds per acre in the first example and 65 pounds per acre in the second. Recall from chapter 2 that the FWS and the PHS tested for potential wildlife effects using concentrations of DDT at 5 pounds per acre or less (and typically less than 2 pounds per acre and often half of 1 pound per acre). Bishopp's statement signaled profound disparities between recommended rates of application and actual rates. Even if the high rates of application were somehow justified, the sheer volume of DDT applied (4.3 million pounds on tomatoes in California alone) boggles the mind. Unfortunately, these are the only two instances in which Bishopp cited the actual application rate of DDT. In numerous other examples, he noted the monetary savings in millions of dollars or dollars of cattle per pennies of DDT.

In addition to asserting the considerable benefit of DDT to meeting the nation's growing demand for food production, Bishopp also noted

141

WHAT'S THE RISK?

the demonstrated benefit of the chemical in the fight against infectious diseases, namely insect-borne disease: "The development of DDT and other new insecticides for controlling disease-carrying insects represents one of the most important advances in medical history. The control of malaria, typhus, encephalitis, dengue fever, yellow fever, filariasis, and other diseases has improved the health of man and increased his life expectancy throughout the world."[82] Bishopp noted that malaria, which he called the "most important disease of man in the world," could be effectively and economically controlled by spraying residual DDT, BHC, or chlordane in homes. Statistics from the World Health Organization supported this claim, and Bishopp concluded by noting, "The almost complete elimination of malaria from the U.S. was hastened by the spraying of 800,000 homes with DDT during 1950 by the U.S.P.H.S. and the State Health departments."[83] Bishopp could enumerate the benefits of DDT to agriculture and public health at great length, and it is clear that such benefits were profound.

And yet Kleinfeld intended to clarify the risks of DDT as well. To that end, Kleinfeld focused on USDA's recommendations regarding DDT residues in milk and cited the following paragraph from a 1945 paper in *Science:* "These preliminary observations prove that with continued oral administration of DDT to goats and rats, there is eliminated in their milk a toxic substance which produces symptoms indistinguishable from DDT intoxication. The data strongly suggest the need for more intensive research on the toxicity of milk from dairy cows ingesting DDT residues either from sprayed or dusted forage plants or from licking themselves after being sprayed or dusted with this insecticide."[84] This statement contradicted one of Bishopp's statements, and Kleinfeld asked Bishopp whether it was safe to assume that DDT applied as a dust or wettable powder in water would not be absorbed by cattle. In responding, Bishopp noted that the *Science* paper was based on ingestion or more specifically experimental feeding and that the USDA recommendations noted that there was a hazard in connection with feeding crops that carried DDT to dairy stock in any considerable amount. When pressed, Bishopp elaborated on this point: "There was no reason to assume that the spraying of barns, that is, putting down a persistent residue, on the walls and ceiling or spraying the cattle with the wettable

142

WHAT'S THE RISK?

powder, which is really just technical DDT, should result in any con-tamination of the milk."[85]

Kleinfeld next asked Bishopp to expand on his statement that the Oklahoma Experimental Station had found that dairy cows sprayed with DDT excreted the chemical in their milk. In this case and many others, Bishopp deflected the question to Edward F. Knipling, also of the BEPQ in Orlando (see chapter 2). Knipling explained that the work in Okla-homa involved what he called "excessive doses of DDT," which greatly exceeded the recommended doses for insect control on dairy cattle. The BEPQ replicated the studies using DDT as recommended and generally used. They reported the results of this research at the Texas Entomologi-cal Society meeting in February 1947, and *Agricultural Chemicals*, the trade journal of the National Agricultural Chemicals Association, re-viewed the paper in its April 1947 issue. Knipling noted that the research at the Texas Experimental Station did not merit an official release, given that the public consumption of DDT would not exceed 0.25 ppm.[86]

Kleinfeld next turned to the USDA's official publication, "The New Insecticides for Controlling External Parasites of Livestock" dated April 1949, which recommended against DDT application to dairy animals producing milk for human consumption due to the appearance of the chemical in milk at potentially hazardous levels as judged by the FDA. Nor should DDT be used in places where milk could be contaminated by the chemical.[87] When Kleinfeld asked Bishopp if he believed the statement to be a sufficient and direct recommendation to dairy farmers not to use DDT as described in the statement, Bishopp deflected the answer to his colleague Knipling, noting that the USDA issued more direct statements to farmers within circulars. At this point, Miller de-manded clarification: "Well, I took it from your testimony this morning that you think there is little or no harm that comes from the use of DDT around dairy barns. Am I correct in that assumption?"[88] Bishopp acknowledged that he ought to qualify his answer. Again, Miller pressed him to be clear, and Bishopp responded: "We are definitely recommend-ing against the use of DDT in dairy barns and on dairy cattle, dairy plants, milk houses, and all such places as that."[89] This statement ad-dressed Miller's concern that Bishopp on behalf of the USDA was de-fending the use of DDT in dairy barns, but Knipling seemed to feel that

WHAT'S THE RISK?

the issue still required clarification, and his comments shed light on a fundamental fracture in the history of toxicological regulation.[90] Namely, the USDA could and did conduct research on potential pathways for contamination of milk and other agricultural products, but it was the responsibility of the FDA to determine the levels of contamination that would be hazardous to humans.[91]

The cloud of uncertainty surrounding the toxicity of DDT and other chlorinated hydrocarbons did not extend to other chemical insecticides, for example, the organophosphates. Kleinfeld examined Bishopp on the use and toxicity of parathion. Bishopp rejected the notion that the chemical was widely used, noting that its use was more or less restricted to certain extensive crops, such as wheat, and specific infestations, including green bugs and fruit insects. Typically, the USDA recommended the use of parathion only in cases where less hazardous chemicals failed to control insects. Kleinfeld sought to establish on the record that parathion was extremely toxic so he pressed Bishopp with a series of questions regarding the toxicity of the chemical.[92] Despite recognizing the risks, including eight deaths and forty-eight cases of severe toxemia, Bishopp admitted that the USDA had not recommended against the use of parathion on fruits.

When Kleinfeld asked Bishopp to provide a safe residue level for any one item of the typical American diet for parathion, Bishopp initially deflected the question to the FDA, but Kleinfeld persisted in soliciting his opinion. Bishopp replied that he believed the FDA published a statement that 2 ppm may be safe, but that he felt that level was a *bit high,* noting that toxicologists pointed out that parathion metabolized readily.[93] Even in the case of parathion, one of the most toxic insecticides ever to reach general use, Bishopp noted that its toxicity could be mitigated by its quick rate of metabolism. Kleinfeld cited a paper, "Absorption of DDT and Parathion by Fruits," presented at the 1949 meeting of the American Chemical Society that found parathion in the peel but not the pulp of harvested oranges, lemons, and grapefruit. Based on the weight of the peel, 3 to 5 ppm of parathion were found in the peel of Valencia oranges six months after treatment with standard dosage. Bishopp noted that Valencia oranges have "pretty thick skins," but Kleinfeld countered by asking whether orange peel was sometimes candied and also used in animal feed. Bishopp acknowledged both potential path-

144

WHAT'S THE RISK?

ways of exposure but argued that the parathion was pretty largely destroyed in those products between harvesting, processing, and consumption. Kleinfeld again pointed out that parathion was present in quantities of 3 to 5 ppm up to six months after treatment, which Bishopp acknowledged was a considerable amount of time. Despite Kleinfeld's pointed questions, Bishopp's testimony did little to clarify for the committee whether parathion actually posed risks to consumers even as Bishopp recognized the chemical as one of the most toxic in use with a safe residue level of 2 ppm (or less, in his opinion).

Even Arnold J. Lehman, director of the Division of Pharmacology at the FDA, could not clarify the risks associated with parathion. Lehman was present during the early days of the hearings, perhaps while he waited to be called to testify. Repeatedly, congressmen called on him to address aspects of chemistry and toxicology. During Lehman's formal testimony, Kleinfeld asked him to describe parathion and its use as an insecticide. Lehman responded: "From my own standpoint, having the interest of the consumer in mind, parathion is probably a safer insecticide than DDT."[94] When Kleinfeld asked if he knew of reports of fatal poisonings, Lehman reported that there had been nine fatal cases of poisoning with parathion. Thomas G. Abernathy (Mississippi) focused the line of inquiry to whether parathion was harmful to a crop and to consumers after eating the crop. Lehman responded definitively: "I think I can answer the question. Parathion is a liquid. It penetrates the skin. It is very poisonous. Very small amounts will produce fatal poisoning."[95] Nevertheless, when Abernathy pressed the point and asked again if parathion could be used on crops, Lehman replied: "I think that it is safe for use."[96] Again Abernathy asked if any damage would result to the consumer in eating a crop on which parathion was used. Lehman responded: "No. There is no evidence that I know of."[97] At that point, Miller noted that he had asked his wife what she used on ants in their home, and he discovered some parathion sprays. He refocused Abernathy's question: "Do you think that [it's] harmful in spraying, as women do, all over the country, to get rid of insects. Is there danger in using parathion?" Lehman answered, "There is." He added a facetious comment: "I hope Mrs. Miller is an expert in the use of insecticides"[98] Miller noted that she was not and took the discussion off the record. When testimony resumed, they did not return to the toxicity of parathion.

145

WHAT'S THE RISK?

Lehman did not clarify why he believed that a highly toxic chemical like parathion was safe for use on crops and safe for people who ate the foods produced from such crops. Based on his publications, Lehman believed that parathion's rapid decomposition would protect crops and consumers alike (see chapter 4).

In order to sort out the different views on the toxicity of insecticides, Kleinfeld sometimes introduced previous testimony into his questions. In the case of chlordane, a chlorinated hydrocarbon like DDT, Kleinfeld cited Lehman's testimony in which he had testified that chlordane was at least "four times as toxic as DDT."[99] Lehman called chlordane "one of the most toxic of insecticides we have to deal with."[100] Chlordane, in Lehman's opinion, had no place in the food industry where the possibility of contamination existed. Nor was the chemical appropriate as a household spray or in floor waxes. When Kleinfeld asked Bishopp for the recommended uses of chlordane according to the USDA, Bishopp read a lengthy statement by Ralph Heal, technical director of the National Pest Control Association. Heal argued that the pesticide control industry had used chlordane extensively and that only one employee had shown a sensitivity or allergic reaction in one particularly large firm with more than one hundred employees who spent roughly half their time applying chlordane. Such a record contrasted even with one of their least toxic insecticides, pyrethrum, to which 14 percent of employees developed an allergic response. Bishopp's response to Kleinfeld offers a glimpse of the considerable extent to which USDA had become co-opted by the chemical companies. Historian Pete Daniel has extensively documented regulatory capture of the USDA, specifically the agency's clearinghouse for pesticide approval, the Agriculture Research Service, by the pesticide manufacturers (see below).[101] Bishopp's statement, however, represents an early example of capture and a frank acknowledgment of industry ties.

It is unclear the extent to which Bishopp's views prevailed among growers and, significantly, producers of food. One clear statement of dissent came from Dr. L. G. Cox representing Beech-Nut Packing Company. In his prepared statement, Cox noted the problems that Beech-Nut faced with the advent of synthetic insecticides like DDT. As early as 1947, Beech-Nut management undertook an extensive and expensive research program on pesticides residues. As a result of this research,

146

WHAT'S THE RISK?

Beech-Nut adopted a "near zero tolerance level" in baby foods. Citing the research of Fitzhugh and Lehman and the recommendations of the American Medical Association, among others, Cox noted that producing baby food necessitated caution: "Since we are manufacturers of baby food, we have had to take into account all those factors—prenatal, environmental, physiological, and structural—which may cause a baby to react to food residues in a manner different from the adult."[102] Such a statement reflects a sophisticated understanding of environmental exposures and dose-response. Cox's statement suggests that infants were particularly vulnerable to environmental exposures. Acting on this concern, Beech-Nut adopted precautionary measures to ensure the safety of their products. Of particular concern was the lack of adequate analytical methods of residue analysis. Cox noted that Beech-Nut had good analytical methods for DDT, DDD, BHC, parathion, methoxychlor, and a few other insecticides not in general use. However, satisfactory analytical methods were lacking for chlordane, toxaphene, aldrin, dieldrin, and heptachlor. In light of this situation, Beech-Nut established an elaborate field survey program through its agricultural purchasing department, which the company enforced through a growers' contract prohibiting the use of certain insecticides. Cox cited Lehman's opinion that chlordane had no place in the food industry, in sharp contrast to Bishopp's recommendations on behalf of the USDA.

Contamination of crops with pesticides of known and indeterminate toxicity represented just one of the serious problems that Beech-Nut confronted. Certain pesticides corrupted the flavor of foods. BHC posed particular problems, and Cox cited separate cases in which the company rejected squash, peaches, celery, spinach, sweet potatoes, apples, and peanuts. The problem of off-flavors was particularly acute in peanuts. Beech-Nut researchers noted a degree of off-flavor in peanut butter containing 2 ppm of BHC after a five-month shelf life. Chlordane and lindane also caused off-flavoring.[103]

Costs of research and quality control, procurement, processing losses, and equipment had amounted to an average of more than $110,000 per year for the five previous years. Such costs inspired seven specific recommendations, including evidence of adequate toxicological testing, development of suitable pesticides residue analysis, information on the approximate range of residues on produce, guidelines for removing

147

WHAT'S THE RISK?

pesticide residues, a requirement that the FDA establish and publish a tentative tolerance in the *Federal Register* within ninety days following registration, FDA sample analysis following the publication of the tolerance, and a requirement that if tolerances were exceeded after a ninety-day warning period, the tentative tolerances would be made official and seizure proceedings instituted.[104] If anything, Cox's statement on behalf of Beech-Nut revealed that one food producer desperately sought oversight of the new chemicals that had proliferated on farms across America.[105] Moreover, Cox revealed that Beech-Nut's in-house scientific group had a sophisticated awareness of environmental exposures and the particular vulnerability of infants, which drove them to advocate precaution with respect to residues of the new insecticides. In both respects, Beech-Nut seems to have been quite progressive in its outlook and its willingness to underwrite precaution. Unfortunately, Beech-Nut's precautionary approach to the new insecticides, as presented by Cox, seems to have been a single exception to the general incorporation of pesticides by food producers and distributors in America.

After hearing and questioning expert witnesses from industry, USDA, FDA, PHS, and other organizations, members of Congress could track several themes. Again and again, they heard that synthetic insecticides enabled farmers to produce crops and thus feed a growing nation. Scientific opinion on the risks of the new insecticides varied widely. On one hand, some scientists could point to specific problems with DDT and other chlorinated hydrocarbons, such as the development of resistance in certain insects and transfer of DDT from cows to calves, which led to advisories against its use in dairy barns. On the other, specific cases of poisoning in humans were rare. Toxicological analysis focused on symptoms of acute poisoning, but the chronic toxicity of DDT remained shrouded in scientific uncertainty. Notably, the leading authority on environmental cancer argued that not enough time had passed to determine if chlorinated hydrocarbons caused cancer in humans. The considerable toxicity of organic phosphates, however, provided a much sharper image of risk. Even as the BEPQ assistant chief questioned whether the tolerance for parathion should be 2 ppm or lower, he left no doubt that it was one of the most toxic chemicals known to mankind. Yet, unlike persistent chlorinated hydrocarbons, it metabolized quickly in the environment. Scientific uncertainty and sharp distinctions be-

148

WHAT'S THE RISK?

tween science and policy provided few clear pathways for further regulation and often left members of Congress confused and exasperated.
Nevertheless, political scientist Christopher Bosso has argued that the
hearings established two major points for those seeking legislative reform: "(1) food and chemical industries did not have consumer health as
their primary orientation, and (2) the FDA had no mechanism for know-
ing beforehand which chemicals reached the consumer, and with what
effects."[106] As chairman of the committee, Delaney concluded the hearings with his observation that committee members on the whole supported premarket testing of food additives and strengthening of FDA
regulatory power.

In the aftermath of the Delaney Hearings, Congress passed two major amendments to the FFDCA. The first was adopted by Congress in
1954 and it became known as the Miller Amendment (§ 408), named for
Miller himself. As we have seen, certain legislators and regulators worried that feeding a growing American population required chemical insecticides, while others were concerned about the potential risks that
these chemicals, particularly chlorinated hydrocarbons and organophosphates, posed to consumers. The Miller Amendment addressed public
health concerns by allowing registration only if manufacturers presented
data that demonstrated residue levels on food crops posed no danger to
public health. This stipulation granted the FDA the authority to set tolerances for each pesticide and crop, specifically raw agricultural com-
modities, such as fresh fruits, vegetables, or milk.[107] Yet the amendment
limited FDA's regulatory jurisdiction to pesticides used on foods; non-
food uses remained under the USDA, even in cases where pesticides
could contaminate the environment and eventually produce exposures
in foodstuffs. Historian John Perkins argued that the Miller Amendment legitimated the use of insecticides by establishing "insignificant"
legal doses. According to Bosso, "Congress thus assured the public that,
while residues might remain in or on food, the levels were not sufficiently toxic to warrant concern."[108] Under this amendment, the FDA
needed to weigh the benefits of insecticides against the risk. This state
of affairs prompted environmental scientist John Wargo to argue, "This
dual standard, necessitating the protection of both public and economic
health, has become the essence of the nation's pesticide control strategy,
as structured by FIFRA and FFDCA."[109]

WHAT'S THE RISK?

In 1958 Congress approved the insertion of the Delaney Clause, which stated: "No additive shall be deemed safe if it is found to induce cancer when ingested by man or animal," within the general safety clause of the FFDCA (§ 409).[110] The Delaney Clause effectively prohibited the FDA, which set food tolerances, from approving any food additive shown to induce cancer in animals or humans. If a food additive (for example, an insecticide) caused cancer, the policy prohibited registration or set the tolerance for approved uses to zero, in effect banning the food additive. The regulatory power of this clause was not lost on legislators who supported the farm bloc: the FDA must ban any suspected carcinogen. Wargo has noted that in passing this legislation, Congress worried that this "zero risk" standard conflicted with the risk-benefit standard that balanced potential risks (toxic residues) with potential benefits (enhanced crop production).[111] According to the environmental historian Nancy Langston, the Delaney Clause had the potential to be revolutionary by stipulating that "any substance known to cause cancer in test animals could not be added to food in any quantity whatsoever."[112] Thus, in Langston's view, the Delaney Clause was a formal, legal expression of precaution, but its focus on carcinogens distracted attention from other hazards.

The first application of the Delaney Clause followed shortly after President Dwight D. Eisenhower signed it into law in November 1958.[113] Barely one year later, HEW Secretary Arthur S. Flemming invoked the clause to advise consumers not to buy cranberries, much to the consternation of growers who objected to the negative publicity so close to Thanksgiving (when 70 percent of cranberry sales occur). Flemming noted that some cranberry products on the market had been contaminated with aminotriazole, an herbicide with USDA approval. In 1957 the FDA issued a warning that aminotriazole left unsafe residues on fruit, and it seized and froze contaminated berries for further analysis. Tests for carcinogenicity took two years and they revealed that aminotriazole caused tumors in rats. These findings precipitated Flemming's announcement and one of the first cases of something approaching mass hysteria as some states banned cranberries or urged caution. Growers and the USDA attacked Flemming and the FDA, arguing that they had not been consulted nor given adequate notice of the announcement. Desperate countermeasures to reestablish consumer confidence included

150

WHAT'S THE RISK?

televised broadcasts of Vice President Nixon and other public figures eating generous portions of cranberry sauce during their Thanksgiving feasts. Despite such efforts, cranberry sales plummeted, and losses to the industry rose to a staggering fifteen to twenty million dollars during 1959 (growers found not to be at fault later recovered ten million dollars in indemnity payments from the USDA). Despite the rather conflicted results of the first application of the Delaney Clause, the hearings and resulting legislation served as a crucial (albeit partial) step in the regulation of chemical pesticides.

As we have seen, Congress, in the passage of FIFRA and the prolonged discussions of risks associated with chemical pesticides during the Delaney Hearings, had already begun to evaluate the testimony of toxicologists both in and out of government. Scientific uncertainty and lack of familiarity with the long-term effects of the newer insecticides stymied legislators' efforts to draw general conclusions regarding risks associated with insecticides. Legislators were asking the questions that needed to be asked, but concrete answers remained elusive as the opinions of experts from the USDA, the FDA, industry, and academia diverged widely. Particularly unclear were the long-term risks associated with DDT and other chlorinated hydrocarbons. Certain target insects developed resistance to DDT, and evidence of transfer from cow to calf prompted USDA advisories regarding the use of DDT in dairy barns as early as 1949. One expert after another acknowledged the dangers of acute poisoning yet wondered about chronic toxicity. Evidence of risk to humans in particular remained murky. Even the nation's leading pro-environmental cancer authority argued that not enough time had passed to determine the tumor-producing potential of DDT in humans. There was no question as to the toxicity and substantial risk of organophosphates, but many believed that their rapid decay offset risks. In striking contrast to most growers and food producers, officials at Beech-Nut Foods adopted a precautionary approach and tested for pesticide residues in its foods.

Despite the uncertainty about pesticides, Congress revised existing legislation to address potential risks. The Miller Amendment restricted residues in foods and empowered the FDA to establish tolerances for foodstuffs. Seeking to minimize the risk of cancer through pesticides, Congress approved the Delaney Clause, stipulating that the FDA set the

151

WHAT'S THE RISK?

tolerance for any carcinogenic food additive to zero and effectively prohibiting FDA from registering any food additive shown to be carcinogenic (including pesticides). The first test of the new legislation came when the FDA advised consumers not to purchase cranberries after some cranberry products revealed contamination by aminotriazole. As the market for cranberries collapsed in the midst of mounting consumer panic, the president and vice president attempted to allay fears and growers defended their product. The cranberry scare brought the problems of chemical food additives and the new legislation to national attention.

CHAPTER 6

Rereading *Silent Spring*

By the late 1950s, toxicologists at the Tox Lab, the FDA, and elsewhere were working to establish toxicology as an independent discipline through courses, textbooks, a professional society, and a scholarly journal. Most Americans had little access to scientific research or the debates in the halls of government. As in the past, it again fell to popular science writers to bridge the gap between scientists, policy makers, and the public. Notable among them was Rachel Carson, whose *Silent Spring* quickly became a bestseller. *Silent Spring* alerted Americans to the hazards of insecticides, but it also inspired renewed interest within government even at the executive level. President John F. Kennedy called for further study through the President's Scientific Advisory Committee, which recommended that Congress review interagency coordination. During the resulting hearings, Congress called on now-familiar witnesses to clarify risks associated with pesticides. Where testimony revealed obvious deficiencies in existing legislation, Congress took decisive action, but the strands of pesticide risk and benefit remained entangled in a Gordian knot.

In August 1958, Kenneth DuBois sought to expand the research and teaching program in toxicology at the University of Chicago. The Toxicity Laboratory had been renamed the U.S. Air Force Radiation Laboratory in 1953, and all funding ($90,000) was dedicated to radiation research. This research extended well beyond Geiling's work on radioisotopes in pharmacology. For example, Doull later recalled that one of his first tasks under the new contract was to collaborate with two other researchers to establish a screening program for radio-protective agents.

153

REREADING *SILENT SPRING*

Critically, the research group obtained LD_{50}s for mice on several thousand agents. As Doull remembered: "The resulting large data base of acute toxicity data in male mice has subsequently proven to be of more lasting value than the few radio-protectors we found."[1]

Although by 1958 the budget had expanded to $275,000, of which approximately half went to research and graduate training in toxicology, DuBois believed that a distinct toxicology laboratory would greatly enhance the Division of Biological Sciences at the university. His justification for such a program was that the discipline of toxicology had developed to such a degree that designated lectures in related courses on pharmacology could no longer encompass the breadth and depth of the burgeoning field: "However, the tremendous increase in the use of chemical agents for industrial, agricultural and household purposes, and the anticipated widespread use of atomic energy have introduced many toxicological problems. The field is expanding so rapidly that it is no longer possible to include adequate coverage of the subject in the formal courses of related disciplines nor for the teachers of other disciplines to keep abreast of developments in the field."[2] DuBois noted that at least one other university, the University of Rochester, was initiating a program in toxicology. Unfortunately, the administration at the University of Chicago did not approve DuBois's proposal as the dean of biological sciences supported molecular approaches to the study of biology over whole animal studies.[3] The shift to molecular approaches at Chicago reflected a trend at many of the leading universities in the U.S. But what appeared to be a curse for the Tox Lab proved to be a blessing for the emerging field of toxicology: many of the Chicago toxicologists found positions at other universities where they developed programs that conformed to the model established in Chicago. Doull, for example, moved to the pharmacology and toxicology program at Kansas University Medical Center in 1967.[4]

During the 1950s, when more and more departments of pharmacology were introducing courses devoted to toxicology, DuBois was writing the *Textbook of Toxicology* with E. M. K. Geiling, who had retired from the University of Chicago and moved to Washington, D.C., to join the Division of Pharmacology at the FDA. Published in 1959, the *Textbook of Toxicology* was the first of its kind in the United States. In its preface, DuBois and Geiling repeated verbatim DuBois's argument for

154

REREADING *SILENT SPRING*

establishing departments of toxicology. In the text, they covered the subject of toxicology in fifteen chapters, emphasizing classes of toxic agents, for example, air-borne poisons, metals, radiation hazards, pesticides, and household poisons. The *Textbook* also treated the historical development, general principles, and medico-legal aspects of toxicology.

In his chapter on pesticides, DuBois reviewed the "toxicity, pharmacological effects, and methods of prevention and treatment of accidental poisoning by the pesticides of greatest importance and usefulness at the time."[5] He divided the insecticides into three groups: the inorganic, the synthetic organic, and the botanical insecticides. DuBois neatly summarized the argument for and the challenges posed by the proliferation of insecticides.[6] As a summary of risks and benefits, DuBois addressed many of the medical, economic, and biological aspects of insecticides, noting that the possibility of developing resistance necessitated a variety of insecticidal material, much like antibiotics. In describing the two major classes of synthetic insecticides, DuBois argued that individual members within each class of compounds "exhibit differences in chemical configuration and physiological actions, but their effects on mammals are essentially the same from the clinical standpoint and the compounds can, therefore, be described as a group."[7] Such thinking would eventually become very important from a regulatory, as well as a toxicological, perspective.[8]

When they established the journal *Toxicology and Applied Pharmacology* in 1959, the inaugural editors—Frederick Coulston, Arnold J. Lehman, and Harry W. Hays—hoped that it would stimulate investigators to publish extensive toxicological studies, that it would provide an outlet for papers by students being trained in toxicology, and that it would serve to centralize important toxicological research which would in turn facilitate the work of the investigators.[9] Like Geiling and DuBois, who became the managing editor of the journal in 1960, Coulston, Lehman, and Hays believed that toxicology needed to establish its independence as a discipline because toxicologists had greater responsibilities and thus needed specific training in toxicology, echoing views expressed by Dubois and Geiling in their textbook.[10] Toxicology was such a broad and significant topic of study that it required disciplinary status. Doull recalled another reason for the establishment of *Toxicology and Applied Pharmacology*: "The reluctance of the *Journal of Pharmacology and*

REREADING *SILENT SPRING*

Experimental Therapeutics to publish tox studies on products or chemicals was not mentioned [in the editors' preface] although it was widely recognized and was certainly one of the reasons for creating the new tox journal."[11] It seems appropriate that the very first paper published in *Toxicology and Applied Pharmacology* addressed the toxicity of organic phosphates: "The Subacute Toxicity of Four Organic Phosphates to Dogs" by Martin W. Williams, Henry N. Fuyat, and O. Garth Fitzhugh (FDA pharmacologists who had conducted many of the original studies on DDT).[12]

Little more than a year later, in 1961, 9 toxicologists met to discuss forming a society devoted to toxicology. It should not be a surprise that Lehman attended (DuBois could not due to illness). A few academics and in-house toxicologists for chemical concerns also attended.[13] They became the founding members of the Society of Toxicology. Some of the founders were concerned about the impact the new society might have on the Society of Pharmacology (the existing body that addressed aspects of toxicology). Still, they proceeded with their plans and elected Arnold Lehman as honorary president, Harold Hodge as president, and DuBois as vice president. By the time the first meeting was held (in Atlantic City in 1962), there were 183 charter members. The new society elected Torald Sollman, von Oettingen, and Geiling as honorary members.

Geiling, in particular, assisted the new society by advising the founders on how they could distinguish themselves from the Society of Pharmacology. In this role, he drew on his association with J. J. Abel, who had endeavored to disentangle pharmacology from the disciplines of physiology and biochemistry. Doull recalled Geiling's important advice to the society: focus on the unique aspects of the new discipline to separate it from the old while identifying the societal benefits of the new discipline. Describing pharmacology as the science of drugs had distinguished the field from physiology and biochemistry, and therapeutics provided a rationale for its benefit to society. Toxicology could be defined as the science of poisons to separate it from pharmacology, and safety evaluation would justify public support.[14] According to Doull, Geiling believed that the society should accentuate its contribution to society and the greater good. From the outset, Geiling believed that the Society of Toxicology should stress the applications of toxicological research. Likewise, including the words "applied pharmacology" in the

REREADING *SILENT SPRING*

title of the new journal indicated a new commitment to applied research and distinguished its concerns from the pure research of the *Journal of Pharmacology and Experimental Therapeutics* (the main existing vehicle for publishing research reports).

Geiling further expanded on the importance of clear definitions and boundaries: "When we define toxicology simply as the adverse effects of chemicals on living systems without including the use of that information to evaluate safety or predict risk we describe what we do but not why we do it. If our discipline focuses on this limited mission, *we risk eroding public support for toxicology, the regulatory process and science in general.*"[15] In the very act of establishing toxicology as an academic discipline, Geiling (and the other founding and charter members of the Society of Toxicology) stressed the importance of public support. Such a preoccupation with public support seems out of place for an academic discipline, which by its very nature ought to be independent of public opinion. Perhaps Geiling, with his long association with crises that subjected science to public scrutiny dating to the Elixir Sulfanilamide tragedy, appreciated the ongoing responsibility of toxicology to regulators and the public. Although the toxicologists may have recognized the importance of public support, they left it to popular science writers to interpret toxicology's findings for laypeople.

The consolidation of toxicology as a discipline among government and academic researchers had little impact on popular conceptions of changes in the natural world. Still, important insights could be gleaned from careful study of the growing body of toxicological literature. Several science writers simultaneously took up the subject of environmental contamination by pesticides, and it was these authors who educated the public. The best known of these writers was Rachel Carson, whose *Edge of the Sea* and *Under the Sea Wind* had informed countless Americans about the intricacies of the natural history and ecology of the seashore and oceans. There were, however, at least two other science writers examining what was known about environmental chemicals: Robert Rudd, a professor at the University of California, and Lewis Herber, who, like Carson, was a freelance writer.[16]

In *Silent Spring,* published in 1962, Carson established a hierarchy of insecticides. She first took up the chlorinated hydrocarbons, starting with DDT, and progressively described other chemicals in the class,

REREADING *SILENT SPRING*

Rachel Carson, photograph by Richard Hartmann,
Courtesy of Magnum Photos.

including chlordane, heptachlor, dieldrin, aldrin, and endrin. Carson
wove details about their toxicity to mammals, birds, and fish into her
descriptions of the chlorinated hydrocarbons. In just a few pages, Carson introduced such concepts as bioaccumulation, lipofelicity (the bond-
ing of chemicals to fats), passage of chemicals from mother to offspring
via breast milk, and liver toxicity, all of which occurred at the residual
levels found in food. Among the chlorinated hydrocarbons, she identi-

fied endrin as particularly toxic: five times more toxic than dieldrin and many times more toxic than DDT.[17] Nevertheless, Carson did not believe that chlorinated hydrocarbons posed the greatest threat to humans and wildlife: she had yet to address the organic phosphates.

Carson left no doubt where organic phosphates stood in the hierarchy of insecticides: "The second major group of insecticides, the alkyl or organic phosphates, are among the most poisonous chemicals in the world."[18] Carson went on to describe ironically the development of the organic phosphates as nerve gases during World War II and the incidental discovery of insecticidal properties; but it is her powerful description of the major effect of the organic phosphates on organisms, insects and warm-blooded animals alike, that sets her account apart from previous reports.

Aware that her subject demanded precision, Carson described the normal function of the central nervous system in detail, including the critical role of a "chemical transmitter": acetylcholine, which under normal conditions facilitated passage of nerve impulses and then disappeared. Excess acetylcholine or its continued presence could wreak havoc on the central nervous system, leading to tremors, muscular spasms, convulsions, and death. Carson proceeded to describe the critical role of cholinesterase in ensuring that the body never built up a dangerous amount of acetylcholine. By inhibiting cholinesterase, organophosphate insecticides disrupted this process: "But on contact with the organic phosphorus insecticides, the protective enzyme is destroyed, and as the quantity of the enzyme is reduced that of the transmitting chemical builds up. In this effect, the organic phosphorus compounds resemble the alkaloid poison muscarine found in a poisonous mushroom, the fly amanita."[19] Thus Carson elucidated the relation between the symptomology of cholinesterase inhibition and the normal function of the nervous system in a way that made clear the risk organophosphate insecticides, such as parathion, posed to humans.

But what was the risk to people who were not exposed on a regular basis? Carson answered this question with additional data showing that seven million pounds of parathion was applied in the United States and the amount used on California farms alone could "provide a lethal dose for 5 to 10 times the whole world's population."[20] What saved the people of the world was the rate at which the organophosphate chemicals decomposed, as we have already seen. They broke down into harmless components rapidly, in comparison to the chlorinated hydrocarbons,

REREADING *SILENT SPRING*

and their residues did not remain as long. Yet Carson challenged this view citing a case in which parathion posed a real threat to workers weeks after spraying: "The grove had been sprayed with parathion some two and a half weeks earlier; the residues that reduced [eleven out of thirty men picking oranges] to retching, half-blind, semi-conscious misery were sixteen to nineteen days old." Carson noted that similar residues had been found in orange peels six months after the trees had been treated with standard doses. On this point, recall the pointed questions directed to Fred Bishopp and others during the Delaney Hearings (see chapter 5).

Not even malathion, the least toxic of the organophosphate insecticides, escaped Carson's perceptive analysis. Malathion, according to Carson, was almost as familiar to the public as DDT. It was used in gardens, household insecticides, and mosquito spraying. Carson revealed that nearly a million acres of Florida communities had been sprayed with malathion in an attempt to control the Mediterranean fruit fly. She questioned the assumption of many people that they could use malathion freely and without harm. According to Carson, it was only an enzyme in the mammalian liver that rendered malathion "safe," but without the enzyme, an exposed person would receive the full force of the poison.[21]

Citing research on potentiation by the FDA and DuBois, Carson explained that the synergy between two organophosphate chemicals could significantly exacerbate the effects of either or both in that one compound could destroy the enzyme in the liver responsible for the detoxification of another organophosphate. Workers could encounter different organophosphates. She noted that a salad bowl could present a combination of insecticides. Moreover, Carson cited evidence that potentiation was not limited to the organic phosphates. Parathion and malathion intensified the toxicity of certain muscle relaxants, and others (malathion included) dramatically increased the effect of barbiturates.

Carson stressed that the advantages that organophosphates possessed over the chlorinated hydrocarbons, such as rapid decomposition, were significantly offset by the dangers of cholinesterase inhibition and potentiation. Her remarks on the acute toxicity of the various pesticides were only a preamble to her larger case: namely, the long-term risks of pesticides (particularly the chlorinated hydrocarbons) to landscapes, wildlife, and humans. In the remainder of *Silent Spring,* the organo-

160

REREADING *SILENT SPRING*

phosphate insecticides recede to the background. Although Carson thoroughly documented and dramatized the lingering damage to soil, water, flora, and fauna associated with chlorinated hydrocarbons, her research revealed few such problems with the organophosphates. Her one example of the effects of organophosphates on wildlife was typically dramatic. In an attempt to control flocks of blackbirds that fed on corn-fields, a group of farmers engaged a spray plane to spray a river bottom-land with parathion. More than 65,000 red-winged blackbirds (*Agelaius phoenicus*) and European starlings (*Sturnus vulgaris*) died, and Carson wondered how many other animals perished from the acute effects of this universally toxic substance. Had rabbits, raccoons, and opossums suc-cumbed as well? Carson was most concerned, however, about unintended effects on humans, specifically workers and children who were most likely to come into contact with organophosphates.[22]

Most of *Silent Spring* focused on the more subtle chronic effects of chlorinated hydrocarbons. Were any such effects tied to organophos-phate insecticides? To support her claim that there might be, Carson recounted the case of ginger paralysis, a condition brought about when people consumed Jamaica ginger as an alternative to the more expensive medicinal products substituted for liquor during Prohibition. The artifi-cial ginger contained triorthocresyl phosphate, which Carson noted de-stroyed cholinesterase in the same way that parathion did. As we saw in chapter 1 more than 50,000 people were permanently crippled by a pa-ralysis of their legs accompanied by destruction of nerve sheaths and the degeneration of spinal cord cells. Carson compared the effects of organophosphate poisonings to ginger paralysis. Even malathion had induced muscular weakness in chickens and, just as in ginger paralysis, the sheaths of the sciatic and spinal nerves were destroyed. Carson even found evidence that regular exposure to organophosphate insecticides might induce mental disease.[23]

By now it should be clear that Carson believed that the organophos-phates posed an equivalent, if not greater, risk to wildlife and humans than the chlorinated hydrocarbons. When she turned to solutions, Car-son advocated biological control. She cited numerous cases in which natural predators and diseases had been introduced to control insect outbreaks. Judicious use of insecticides played a minor role in Carson's vision of pest control, but they needed to be phased out eventually. An

161

REREADING *SILENT SPRING*

awareness of ecological relationships should guide all endeavors to reduce the depredations of insects and other organisms deemed pests, according to Carson: "Only by taking account of such life forces and by cautiously seeking to guide them into channels favorable to ourselves can we hope to achieve a reasonable accommodation between the insect hordes and ourselves."[24] No chemical insecticide offered a genuine solution: "As crude a weapon as the cave man's club, the chemical barrage has been hurled against the fabric of life. . . . These extraordinary capacities of life have been ignored by the practitioners of chemical control who have brought to their task no 'high-minded orientation,' no humility before the vast forces with which they tamper."[25]

A number of historians and biographers have analyzed the dramatic response to *Silent Spring* on the part of consumers, scientists, industry representatives, and legislators.[26] In general, the response to *Silent Spring* split along predictable lines. Carson found her greatest support from environmental activists like Roland Clement, who presented the book's chief arguments in many presentations to the public and various branches of government. Predictably, chemical companies mounted a savage campaign to discredit Carson and the claims she made in *Silent Spring*. One threatened to bring suit against the *New Yorker* after Carson's articles appeared; William Shawn, longtime editor, allegedly relished the possibility of unexpected publicity for the magazine. Still, some environmental scientists who were apparently impartial distanced themselves and criticized some of Carson's interpretations of the evidence of environmental and human health hazard. One wrote that *Silent Spring* was "full of truths, half-truths, and untruths as far as the wildlife was concerned, and I have nothing to say about the human health thing."[27]

Silent Spring was published in 1962, first as series of excerpts across three issues of the *New Yorker* and in book form the following fall. Despite strong sales, its rapid ascent to the bestseller list, and its selection by the Book of the Month, Rachel Carson and *Silent Spring* did not reach many American households until the evening of April 3, 1963, when CBS Reports aired "The Silent Spring of Rachel Carson," featuring the author herself as well as representatives from the USDA and the chemical industry. The host, Eric Sevareid, introduced Carson, noting that there was a pesticide problem and that Carson favored alternative methods and a gradual phase out of chemical insecticides. Carson then

162

presented her thesis and examples of the hazards of pesticides. Among the many officials who appeared after Carson and who represented the USDA or the chemical industry, the most memorable was Dr. Robert White-Stevens, a scientist with American Cyanamid and the voice for the chemical industry. Dressed in white lab coat and tie, quintessential symbols of medical-scientific authority, White-Stevens forecast a dire future in a world without insecticides: "If man were to faithfully follow the teachings of Miss Carson, we would return to the Dark Ages and the insects and diseases and vermin would once again inherit the earth."[28] Despite, or perhaps because of, White-Stevens's demeanor and dire warning, reviews of the report agreed that Carson's urgent message had captured the minds and hearts of millions of viewers. Indeed, the public was roused from complacency with regard to chemical insecticides. Their concerns reached the highest office in federal government.

On August 29, 1962, President Kennedy fielded the following question at a press conference: "There appears to a growing concern among scientists as to the possibility of dangerous long-range side effects from the use of DDT and other pesticides. Have you considered asking the Department of Agriculture or the Public Health Service to take closer look at this?" The president, who was a regular reader of the *New Yorker* and had read Carson's three articles over the summer, replied: "Yes, and I know they already are. I think particularly, of course, since Miss Carson's book, but they are examining the matter."[29] As a result of the president's comment, the Life Sciences Panel of the President's Scientific Advisory Committee (PSAC) took up a study of pesticide use and associated risks.[30]

Led by Jerome Wiesner, the PSAC prepared a report for the president: "The Use of Pesticides." The report opened with an acknowledgment of some element of risk in modern society which strikes a note of inevitability or perhaps fatalism with respect to "progress": "Advances have always entailed a degree of risk which society must weigh and either accept, or reject, as the price of material progress."[31] Here is another example: "The welfare of an increasing human population requires intensified agriculture. This in turn enables the pests to increase, which necessitates the use of pesticides with their concomitant hazards. It thus seems inevitable that, as the population increases, so do certain hazards."[32] Next, the report recognized spectacular increases in agricultural

163

REREADING *SILENT SPRING*

production and an unprecedented freedom from communicable diseases, including malaria, typhus, and yellow fever. At the same time, arboviruses claimed many lives, with malaria leading the list worldwide. Some of the insect vectors, such as mosquitoes that transmitted malaria, produced populations that were resistant to pesticides. The PSAC argued that Americans had come to require and expect efficient agricultural production, protection of health, and elimination of nuisances.[33]

Against the backdrop of the benefits of pesticides, the PSAC expressed equanimity when it came to potential risks, as in this statement: "Precisely because pesticide chemicals are designed to kill or metabolically upset some living target organism, they are *potentially dangerous* to other living organisms. Most of them are highly toxic *in concentrated amounts*, and in unfortunate instances they have caused illness and death of people and wildlife. Although acute human poisoning is a measurable and, *in some cases*, a significant hazard, it is relatively easy to identify and control by comparison with potential, low-level chronic toxicity which has been observed *in experimental animals*."[34] Significant qualifiers moderate most of the claims included in this statement, but the panel acknowledged the need for a more complete understanding of the properties of pesticides and their long-term impact on biological systems.

Throughout the report, the PSAC noted gaps or deficiencies in the knowledge base regarding the toxicity of pesticides. For example, after reviewing the history of FDA procedures regulating chemicals and pesticides, the panel noted that the FDA commonly set human tolerances at 1/100 of the lowest level that caused effects in the most sensitive test animals whenever data on human toxicity was not available. Nevertheless, the FDA had set tolerances for certain compounds, for example, dieldrin, aldrin, heptachlor, and chlordane, despite that fact that a "no effect" level in animals had not been determined. The no-effect level was critical because it established one toxicological baseline (another was the LD_{50}, see chapter 1). Moreover, the lowest effect level could not be determined accurately without an established no-effect level.[35] The PSAC concluded that in certain instances the experimental evidence was inadequate and recommended continuation of FDA review and reassessment.

Like Carson, the PSAC distinguished between the classes of insecticides and noted the action of organophosphates (namely, cholinesterase inhibition) and the relatively high toxicity of the class to humans: "Most

164

REREADING *SILENT SPRING*

organic phosphorus insecticides have relatively high acute toxicities and have caused many fatal and nonfatal poisonings in man. In cases of poisoning, removal from exposure to the compound usually permits rapid recovery."[36] The committee also noted that most organophosphates degraded rapidly and seldom persisted in the environment, with the significant exception of parathion, which had been found to persist for months in soils and appeared in trace amounts in water drawn from deep wells.[37]

In its final recommendations, the PSAC called for the Department of Health, Education, and Welfare (HEW) to develop a comprehensive data-gathering program to determine the levels of pesticides in workers and the general public. HEW should also cooperate with other departments (USDA and FDA) to develop a network to monitor residue levels in air, water, soil, humans, wildlife, and fish. Total diet studies on chlorinated hydrocarbons initiated by the FDA should expand to include data on organophosphates, herbicides, and carbamates in populated areas where they were widely used. Federal funds would assist states in their monitoring efforts. The PSAC also made a series of recommendations regarding toxicity studies, which included determination of effects on reproduction through at least two generations in at least two species of warm-blooded animals, review of chronic effects on organs of both immature and adult animals, and study of possible synergism and potentiation of commonly used pesticides with commonly used drugs, such as sedatives, tranquilizers, analgesics, antihypertensive agents, and steroid hormones. The panel also recommended expanded research on the toxic effects of pesticides on wild vertebrates and invertebrates, echoing Carson's thesis from *Silent Spring*: "The study of wildlife presents a unique opportunity to discover the effects on the food chain of which each animal is a part, and to determine possible pathways through which accumulated and, in some case, magnified pesticide residues can find their way directly or indirectly to wildlife and to man."[38] On one hand, the President's Science Advisory Committee provided a fairly moderate assessment of the state of knowledge with respect to chemical insecticides and threats to the environment and human health. And yet, on the other hand, the PSAC laid the foundation for the federal response to the apparently ubiquitous pesticides and other chemicals in the environment.

With the release of the report pending, Congress convened hearings to investigate interagency coordination regarding environmental

165

REREADING *SILENT SPRING*

hazards. On May 16, 1963, Senator Abraham Ribicoff of Connecticut called to order the Subcommittee on Reorganization and International Organizations to begin a study of interagency coordination in environmental hazards. Hubert Humphrey, the Minnesota senator and assistant majority leader of the Senate, was designated chair of the subcommittee, but as his additional responsibilities kept him on the Senate floor, Ribicoff presided over the hearing on May 16 and all subsequent sessions. Ribicoff introduced the hearings by stating that they planned to examine the role of the federal government in dealing with contamination of the environment by chemical poisons.[39] Ribicoff went on to note that inasmuch as Americans lived in the space age, they also were in the midst of the chemical age. A consequence of the chemical age included hazards created by the use of chemical poisons to control insects and other pests, eliminate undesired vegetation, and prevent the infection of plants and animals, including humans, by disease organisms. Yet pesticides also yielded benefits, including freedom from disease, more and better food, and great potential in the fight against world hunger, poverty, and disease, according to Ribicoff, who also argued that controlled use must continue. Moreover, knowledge of long-term consequences to human health was inadequate. Ribicoff concluded by reiterating that the purpose of the hearings was to determine the nature of problem; examine the response of government, industry, and the public; and explore the extent to which various federal agencies coordinated their activities and administered their programs.[40]

After a review of the PSAC report and brief questioning of its lead author, Ribicoff began calling witnesses. One of the first to testify was Secretary of Agriculture Orville L. Freeman, who was accompanied by Edward F. Knipling, now the director of the Entomology Research Division at the Agricultural Research Service, and Martin Jacobson, chief scientist of the same division. Freeman quickly asserted the importance of insecticides to modern agriculture. Balancing the benefits of pesticides with acceptable levels of risk was one of the recurring themes of the hearings. During the course of his testimony on May 23, 1963, Freeman underscored the benefits of chemical insecticides: "Chemical pesticides have given us an effective means of protecting our food supply—and if there are certain dangers attendant upon using them, I believe there may be greater dangers in not using them. Without them, the conse-

166

REREADING *SILENT SPRING*

quences in terms of the national diet and the Nation's standards would be serious."[41] Freeman also asserted that with proper controls and safeguards, pesticides could safely be used to protect food, fiber, and forest crops from disease and insect destruction. Finally, he noted that the cumulative effects of minute amounts of various pesticides to humans over the long term remained unknown, but that this lack of scientific evidence suggested the need for further research. Despite his defense of the benefits of pesticides, Freeman acknowledged problems with the registration process, particularly protest registrations (see below), and he described at length efforts at USDA to develop biological controls. Knipling gave a detailed description of control programs using sterile male insects and natural attractants.

As for risks associated with pesticides exposures, Freeman subtly shifted the burden of risk to consumers (and public education campaigns) by noting that the USDA wanted to make the slogan: " 'Use Pesticides with Care—Read the Label' as familiar to the public as Smokey Bear. I believe the public interest stirred by Miss Carson's book, the report of the President's Science Advisory Committee, and these hearing will be of great help in this educational program."[42] At the end of his statement, Freeman questioned the preoccupation with pesticides hazards and reiterated the benefits of their continued use by suggesting potential problems if they were not available. Commercial production of many common fruits (apples, peaches, and cherries) and vegetables (corn, tomatoes, and lima beans) would not be possible or would be greatly reduced. Diseases would affect more vegetables and fruits, their cost would increase, and the amount spent on food for the average family would rise from one-fifth of the average income to one-third. Throughout the hearings, other witnesses emphasized benefits of chemical pesticides.

Despite his firm belief that the benefits of pesticides significantly outweighed the risks, Freeman criticized FIFRA for one provision that subjected the public to danger. On the occasions when the USDA denied the registration of a product, a manufacturer could still register the pesticide "under protest" and distribute it to the public until the USDA had developed the performance and toxicity analysis required to take legal action and remove the product from the market. In Freeman's words, "The Department must carry the burden of proof of establishing in a court of law the danger of a given commodity."[43]

167

REREADING *SILENT SPRING*

Ribicoff clarified Freeman's statements regarding protest registrations. In one example, USDA had denied registration for a device known as a "lindane vaporizer" because it was determined to be too hazardous to be sold for continuous use in homes. However, under protest registration, the manufacturer continued to sell it, and the vaporizer had been implicated in at least one death. Ribicoff wondered why there had been a two-year delay in submitting legislation to correct the problem, but Freeman responded that the matter had been submitted, but no one had acted on it. To place protest registrations into the broader context of pesticide regulation, Freeman explained that of the 55,000 products registered by the USDA since 1947, only 23 had been registered under protest. He wondered if the USDA should be required to carry the burden of proof of harm.[44] Ribicoff did not share Freeman's ambivalence regarding protest registrations and vowed to introduce corrective legislation to Congress, and true to his word, he introduced a bill to amend FIFRA to eliminate registration under protest on May 27, 1963. Ribicoff also demanded that the USDA furnish the names of products registered under protest to the press, noting, "It is a mockery of regulation for the USDA to find a product unsafe and then refuse to tell the public the name of the product."[45]

The most anticipated witnesses, Rachel Carson, appeared before the committee on June 4, 1963. Ribicoff gave her an enthusiastic introduction: "We are dealing with many forces which people say are still mysterious, and it is the purpose of this committee to try to be as constructive as we possibly can, and I think that all people in this country and around the world owe you a debt of gratitude for your writings and for your actions toward making the atmosphere and the environment safe for habitation, not only by human beings but for animals and nature itself."[46] Carson's prepared statement cited examples of ongoing problems and recent findings. Specifically, she reinforced some of her claims regarding the concentration of chemicals by living organisms, noting that oysters concentrated zinc at a level about 170,000 times that in the surrounding water and that marine organisms also concentrated chemicals like DDT. Citing the previous testimony of Secretary of the Interior Stuart Udall, Carson reminded the committee that oysters exposed at levels of only 1 ppb of DDT for one week later contained 182,000 ppb in their tissues, with obvious implications for organisms that consume oysters, including human beings.[47]

168

REREADING *SILENT SPRING*

Lest any doubts remained regarding Carson's recommendations, she enumerated them for the committee. First, all community, state, and federal spraying programs should be required by law to provide advance notice so that all interests could receive hearing prior to any spraying. Second, Carson stressed the need for further medical research and education, quoting the following statement from the PSAC report: "Physicians are generally unaware of the wide distribution of pesticides, their toxicity, and their possible effects on human health."[48] Third, emphasizing the stringent controls on the sale of drugs, Carson urged that the sale and use of pesticides should be restricted, possibly at the state level, to individuals capable of understanding the hazards and of following instructions. Fourth, again citing the PSAC report, and appropriate to the topic of the hearings, she urged that registration require interagency coordination with the participation of the Departments of Health, Education, and Welfare; Interior; and Agriculture. Fifth, citing the compounding of present problems by the "fantastic number of chemical compounds in use as pesticides," she recommended limits on the number of pesticides in use with an ultimate goal of approval for use only when no existing chemical would do the job. Sixth and finally, Carson encouraged support for new methods of pest control that minimized or eliminated the use of chemical insecticides.[49]

Much of the questioning that followed Carson's prepared statement attempted to clarify her position on the relative risk and benefit of chemical insecticides. Ribicoff, for example, raised this point in his first question: "For instance, isn't it fair to say that you are not trying to stop the use of chemical poisons?" Carson acknowledged that this was a fair statement and she agreed that pesticides had produced benefits, but she reiterated that her concern was the serious side effects. Ribicoff elaborated upon his point: "And am I correct, then, that your primary objective is against the indiscriminate use of pesticides and use where they are not necessary, and their excessive use even where they are necessary?"[50] Again, Carson agreed. Others asked for clarification of the major claims of *Silent Spring*, but the most prominent theme was that Carson did not advocate for elimination of chemical insecticides but instead recommended judicious use.

Ribicoff asked Carson to elaborate on her call for additional research into genetic effects at the federal level. Carson felt that the FDA should

169

REREADING *SILENT SPRING*

have a department of genetics or at least a small staff of geneticists to determine genetic effects, which she noted could be quite independent from the toxic effect. Carson's reasoning flowed from her knowledge of medical genetics, which had shown that many human defects and illnesses could result from apparently slight damage to the chromosomes: "But those apparently slight genetic changes cause a whole group of diseases or defects, especially of congenital defects, very often including mental deficiencies."[51] From this point, Carson shifted back to chemical toxicology and the fact that certain chemicals, including some pesticides, also caused chromosomal damage. Juxtaposing these related fields brought Carson to her conclusion: "Now I think those two fields of study ought to be gotten together. We should find out whether the pesticide chemicals in the concentrations in which they are used or at the levels to which they may build up in the human body, are capable of causing these defects and these illnesses."[52] Genetic toxicology was in its infancy when Carson made this recommendation in 1963.[53]

At times Carson called upon the committee to deepen its probe of the pesticides industry and its implications for Americans. For example, Ribicoff noted that many pesticides were intended for home and garden use, but that before the publication of *Silent Spring* individuals had little or no appreciation of their potential dangers. Rather than beginning at the point of use, Carson suggested that the committee might give attention to the type of advertising that introduced consumers to these chemicals. She stated: "I think at the present time and in the past there has been too little to warn the consumer that he is buying and using a very hazardous substance. In fact, the tone of many advertisements of course is quite the contrary."[54] She hoped that the committee would correct this problem of minimizing risk in advertisements. By way of conclusion, Senator James B. Pearson of Kansas summarized Carson's various arguments: "I think from this report and from what you have said that it justifies the work of this committee. We ought not to minimize any dangers, and we should seek an objective analysis, which I have great confidence the chairman will provide. We must not only not overlook dangers, but also must seek to allay any fears the public may have that will come about from this voluminous amount of testimony and hearings that we are conducting."[55]

170

To the extent that *Silent Spring* brought the unintended consequences of chemical insecticides to the attention of Americans, Carson's testimony to the Ribicoff subcommittee on interagency coordination underscored the importance of the soft-spoken biologist in the quest to understand the risks of pesticides and helped to place their benefits in context, much like the CBS report a few weeks prior. Carson succumbed to cancer some ten months after appearing before the committee, but the hearings continued, and to mark the sad occasion Ribicoff read a tribute to Carson: "After Rachel Carson brought her message of concern to the public, it was no longer possible to consider only the benefits of manmade pollution without weighing the risks. It was no longer possible to build a new factory without being concerned about air pollution, a nuclear reactor without being concerned about water pollution, or a new pesticide to make food more abundant without being concerned about the health of wildlife and even people. She was no fanatic trying to wish away the advantages of the 20th century. She was a humanitarian, insisting that man weigh carefully the consequences of his modern technology and strike a balance that will preserve the wonder of nature in the 21st century."[56] Ribicoff concluded: "Today we mourn a great lady. All mankind is in her debt."[57]

Like Carson, both the PSAC and the Committee on Interagency Coordination explored the toxicity of the major classes of insecticides with particular emphasis on the chlorinated hydrocarbons like DDT. The PSAC also noted the risks of organophosphates, such as parathion: "Most organic phosphorus insecticides have relatively high acute toxicities and have caused many fatal and nonfatal poisonings in man. In cases of poisoning, removal from exposure to the compound usually permits rapid recovery. Many of them are degraded rapidly and thus seldom persist in the environment, but some, such as parathion, have persisted for months in soils and have recently been found in trace amounts in water drawn from deep wells."[58] High toxicity and rapid disintegration in the environment were the characteristics that made organophosphates useful as insecticides; high toxicity also made them highly dangerous to use. The PSAC included among its recommendations that the FDA should expand its total diet studies on chlorinated hydrocarbons to include data on organophosphates, herbicides, and carbamates in populated

REREADING *SILENT SPRING*

areas where they are widely used. The PSAC also listed parathion among the commonly used chemicals that the FDA should include in its review of residue tolerances and the experimental studies on which they are based.[59]

Later George Larrick, commisioner of the FDA, elaborated on the agency's efforts to develop meaningful residue tolerances for organophosphates, noting that cholinesterase inhibitors could potentiate when used together. The FDA stopped issuing tolerances in that area and called upon manufacturers to test not only the toxicity of the single pesticide but also the multiple combinations that might be used. Despite increased levels of testing at the FDA, universities, and corporations, Larrick acknowledged the limitations of available information: "The total knowledge, the biological experience and knowledge in this field is not great enough at this time to permit anybody to say with honesty that we know enough about it to be very sure that there are not some synergistic adverse effects possible, and certainly we need to step up research, as Dr. Wiesner said, very extensively in this field to insure a greater protective job."[60]

Not all of the witnesses were as sanguine when it came to the continued use of organophosphate insecticides. Theron G. Randolph, formerly chief of the Allergy Clinic at the University of Michigan Medical School, argued that some individuals became sensitized to certain chemicals as a result of cumulative exposures. This possibility of sensitization led to Randolph's first recommendation (of six): "Persistent insecticides (chlorinated and related hydrocarbons); highly toxic insecticides (organophosphorous material) and biological insecticides which sensitize readily (Pyrethrum) should not be employed in the home or incorporated into other materials for home use."[61] Randolph did not indicate that he was aware that this recommendation would have effectively banned the three major classes of insecticides from household use.

Other researchers cited data that supported Randolph's recommendations. Dr. Irma West, from the Bureau of Occupational Health in the California Department of Public Health placed the risks of organophosphates in the context of other hazardous classes of insecticides: "It is of interest to note that in the past decade in California there have been three different groups of pesticides primarily involved in fatal effects upon humans and wildlife: for children it has been arsenic, for workers organic phosphates, and for wildlife the chlorinated hydrocarbons."[62]

172

REREADING *SILENT SPRING*

To bolster her case against organophosphate insecticides, West cited statistics from the California Bureau of Occupational Health. She interpreted the data as follows: "While organic phosphate pesticides represented about 80 percent of the total reports, they constituted nearly three-quarters of the 268 reports of systemic poisonings, thus indicating the hazardous nature of the organic phosphate pesticides."[63] An analysis of 911 reports of occupational disease attributed to pesticides in California in 1961 revealed that parathion was by far the most frequently reported organic phosphate, accounting for 110 of the 254 reports of occupational illness attributed to organophosphates followed by phosdrin (29 reports) and malathion (24 reports). DDT and other chlorinated hydrocarbons (chlordane, lindane, and keltane) accounted for 34 reports, and the more toxic insecticides in the class (endrin, aldrin, dieldrin, and toxaphene) yielded just 7 reports.

Such data convinced the California Department of Public Health to recommend mandatory medical supervision, including routine cholinesterase tests for all people working with toxic organophosphates. In fact, the state's Department for Industrial Relations adopted the recommendations in their Safety Orders for Agricultural Operations in November 1961. The Safety Orders required employers to engage a licensed physician to provide medical supervision whenever workers were spreading, spraying, dusting, or making other application or formulation of toxic organic phosphate pesticides.[64] The orders further defined medical supervision to include advance planning for prompt care of organophosphate poisoning and cholinesterase determinations or other recognized medical tests before exposure and if necessary after exposure as well.

Not all organophosphates reached the point where they could be marketed to the public. Julius E. Johnson, the manager of Bioproducts for Dow Chemical noted several examples of organophosphates, among others, that were eliminated from further development as a result of toxicological analysis. Dow abandoned several organophosphates after determining that they were highly toxic to mammals via skin absorption or ingestion, with LD_{50}s to rats in the range of one to two mg/kg of body weight.[65]

Nevertheless, organophosphate insecticides posed considerable risks, even in cases in which established guidelines were followed. One epidemic in California involved a recorded ninety-four pickers, but there

173

REREADING *SILENT SPRING*

was no evidence that the recommended fourteen-day interval between spraying and harvesting was violated. Researchers hypothesized that the parathion in use was somehow altered to one or more cholinesterase-inhibiting compounds of considerably greater toxicity, and they suspected paraoxon as the most likely compound. It also seemed likely that the ninety-four pickers, who sought medical attention, reflected a fraction of the total exposed population that would have shown evidence of poisoning, if they had been studied.[66]

When Bert J. Vos, deputy director, Division of Pharmacology at the FDA, appeared before the committee, Ribicoff took the opportunity to return to no-effect level, which the Wiesner Report addressed specifically, noting critically that the FDA had set tolerances for compounds (notably dieldrin, aldrin, heptachlor, and chlordane) even though a no-effect level for animals had never been determined. Ribicoff asked Vos to explain how the FDA had determined tolerances for these and other compounds in the absence of a no-effect level. In response, Vos gave a rambling description of a process that drew on the overall picture of toxicity, comparing the compound to related compounds for which no-effect levels had been determined. Ribicoff asked how he could make such a determination without the no-effect level. Vos attempted to clarify with this statement: "As I said, the tolerance was arrived at by comparing the level at which effect did occur, the severity of the effect, and comparing that with other pesticides which produced similar effects, comparing the levels. The ones at which a no-effect level was not reached, we, you might say, extrapolated on the basis of the way in which the severity effect was at higher levels."[67] For Ribicoff, there was a simpler description of the process: "In other words, what you are doing is guessing?"[68] Vos initially stood firm, but he eventually conceded the point.[69]

Later Ribicoff raised the issue of the no-effect level with several representatives from companies that manufactured chemicals. John P. Frawley, chief toxicologist in the Medical Department of the Hercules Powder Company, and formerly assistant chief of the Chronic Toxicity Branch at the FDA[70], was describing the development of the no-effect level when Ribicoff interjected that if scientists did not perform tests and the chemicals did in fact have genetic effects, was there not the possibility of producing another thalidomide? Frawley answered this question at length by dismissing the comparison with thalidomide since the

174

REREADING *SILENT SPRING*

infamous tranquilizer was used at pharmacologically active levels. For pesticides, scientists relied on the no-effect level: "With pesticides, we have demonstrated the no-effect level by every tool that we can bring to our command today. Then we administer or permit in the environment of man something which is a small fraction of that no-effect level in the animal." Dow's Johnson clarified Frawley's response by noting that for hypnosis, thalidomide had a dosage range between 50 and 200 milligrams per person. For contrast, Johnson cited Hayes, who had said that DDT would amount to 0.284 milligrams per person. The 50-milligram dosage level would be an intake of 200 times as much and the 200-milligram level would be almost 1,000 times as much intake. With this comparison, Johnson hoped to illustrate the difference between an effect level of a drug like thalidomide and a very minute trace level of a pesticide like DDT.[71]

By 1963, as Nancy Langston has shown, thalidomide stood at the center of a major controversy in pharmaceutical regulation.[72] A European company, Chemie Grünenthal, first synthesized thalidomide in 1954 as part of a search for new antibiotics.[73] The new drug revealed no antibiotic properties, but the company patented it and distributed samples to doctors in West Germany and Switzerland. Strangely, Chemie Grünenthal had not developed a scientific protocol to monitor results or to conduct a systematic follow-up, but when patients reported thalidomide's sedative effects, the company discovered a market for the drug. Even before thalidomide was released to the market, there was a report of a baby born without ears after its mother used thalidomide in pregnancy. Nevertheless, on the basis of reports that the drug soothed nausea in pregnancy, Chemie Grünenthal marketed the drug for pregnant women, touting it as the "best drug for pregnant women and nursing mothers," despite a complete lack of studies to consider thalidomide's fetal effects.

When the Richardson-Merrell Company bought the U.S. license for thalidomide in 1960, it submitted a New Drug Application to the FDA. Frances Kelsey, Geiling's former student and colleague at the Tox Lab, had recently taken a position at the FDA, and she found the thalidomide application troubling on two grounds. First, the drug's "curious lack of toxicity" caused her to question the company's safety data. Second, there was an absence of good evidence on its relation to metabolism and excretion. Kelsey rejected Richardson-Merrell's New Drug

175

Application for thalidomide. For months she had to withstand the badgering of the company as it tried to convince Kelsey (and her superiors) to approve the application. In November 1961, reports concerning European mothers who had taken thalidomide during pregnancy began to circulate: their babies had been born with severe birth defects. Kelsey noted cases of children born with hands and feet attached directly to the torso; others were born with limbless trunks; and still others were born with just a head and torso.[74] Kelsey's colleague confirmed the European reports and testified before Congress, which helped Kelsey to block the approval of thalidomide in the U.S. Despite mounting evidence, FDA Commissioner Larrick failed to take action until July 23, 1962, eight months after Richardson-Merrell notified the FDA that thalidomide had been withdrawn from the German market. An investigation revealed that both Chemie Grünenthal and Richardson-Merrell had attempted to deceive regulators. For the part she played in keeping thalidomide off the market in the U.S., Frances Kelsey received the President's Award for Distinguished Federal Civilian Service from President Kennedy. Tragically, tens of thousands of Americans had received thalidomide as Richardson-Merrell had distributed the drug to more than 1,200 physicians for "investigational use." More than 20,000 patients received more than 2.5 million pills during these trials. Most of the exposed women were never notified of the dangers of birth defects.[75]

In drawing a distinction between the cases of thalidomide and pesticides, Frawley and Johnson identified the critical difference: women had been exposed to thalidomide at therapeutic levels, which is to say levels much higher than those found in pesticides, except in extreme cases of acute exposure. Ribicoff redirected the discussion back to the no-effect level by asking if the scientists believed that the product should not go on the market until a no-effect level had been established. Frawley immediately agreed, but Johnson's response was more deliberate, calling for a no-effect level in significant experiments that have a real bearing on the ultimate use of the product. When Ribicoff pressed him for further clarification, Johnson suggested experiments that would indicate such hazards as reproductive effects, or kidney damage, or liver damage. Ribicoff interjected to ask about skin irritation or eye irritation as a potential effect, but Johnson compartmentalized skin irritation as a handling hazard distinct from effects of long-term exposures to minute

REREADING *SILENT SPRING*

dosages. Ribicoff asked two other industry representatives, but neither was a toxicologist by training and both deferred to the opinions voiced by Frawley and Johnson.

Nevertheless, not all representatives of the pesticides industry shared Frawley's and Johnson's view of the no-effect level. Dr. Edmund F. Feichtmeir, manager of product application, Agricultural Research Division at Shell Development Company stolidly evaded and countered questioning regarding the no-effect level from Ribicoff. Ribicoff cited (as Exhibit 126) the definition of the no-effect level (as prepared by FDA). Specifically, the FDA defined the level as follows: "The 'no-effect' level may be defined as that dosage of the test substance which produces no adverse effect as determined by conventional methods."[76] In more technical terms, the FDA related the no-effect level to the LD_{50}: "In the determination of an LD-50, for example, the LD-0 can be considered as the 'no-effect' dose under the conditions of this type of study."[77] In determining a chronic effect, no-effect levels could be assigned to a number of end points: mortality rate in treated versus controls, growth rate in treated versus controls, organ/body weight ratios, inanition, deposition in the tissues, cholinesterase inhibition, histopathology, demyelination, and toxic effect on the fetus. Based on this definition and the statement regarding no-effect levels in the PSAC report (see above), Ribicoff asked Feichtmeir whether it was true that a no-effect level had not been established for dieldrin, one of the stronger chlorinated hydrocarbons sold by Shell. Initially Feichtmeir acknowledged that it was true, but then he proceeded to qualify his statement noting that he was not in a position to say whether it was a no-effect level or not. He concluded that consultants and the FDA had deemed dieldrin safe, and he deferred to their designation.

Ribicoff pursued his line of inquiry pointing out that the FDA was reassessing the question of tolerances and that studies should be made, but the products continued to enter the market even though the no-effect level in animals had not been determined. Feichtmeir reiterated that no-effect levels were a matter of opinion and that the products had been reviewed and tolerances established. He underscored a recent advance in the development of analytical methods: "In the last 2 years we developed what we call an electron capture detector cell as an adaptation to gas-liquid chromatography. By this technique, we can measure in the range of a tenth of a part per billion. . . . I do not know how many of you have an understanding of what this means, but it is about two inches relative to the distance to the moon and back. We have to detect these small differences. With the new tools we can measure blood levels after exposure to the chlorinated materials."[78] Feichtmeir was mistaken in his chronology: James Lovelock invented the electron capture detector in 1957.[79]

Ribicoff refused to be derailed and again asked if Feichtmeir or his company had found a no-effect level for dieldrin, and again Feichtmeir declined to answer directly, noting that it came down to a matter of opinion to be determined by toxicologists, whether Shell's medical consultants or FDA's researchers. Ribicoff reframed the question in light of "good business practice": "Let us say that the National Academy of Sciences determines that there is no no-effect level for Dieldrin. Now, from a good business standpoint alone, forgetting public policy, is there not a duty more or less on you people to determine this before you market the product?"[80] Here Ribicoff introduced the precautionary principle, which, as we have seen, shifted the burden of proof of safety to the pesticide producers from the public and the regulatory network.[81] In response, Feichtmeir argued that Shell had "done that sort of thing when the labels were secured": "In our opinion, there is nothing wrong with these materials at present, based on available data. We think they are safe and can be used, if used properly, without hazard to the public and applied without hazard to the people applying them."[82] Perhaps recognizing the futility of further interrogation, Ribicoff let the subject drop.

Another witness, Ernest J. Jaworski, who was a senior scientist at Monsanto Chemical Company, raised an issue that would later emerge as a topic of considerable contention. Jaworski wondered about the toxicity of natural products. He suggested that toxicologists knew more about the toxicology and pharmacology of pesticides than about many natural products and household products, including food. The Monsanto scientist commented: "We know a tremendous amount about pesticides in general, but I would worry about how little we know in general about food that we eat. What is the chronic effect of eating some given vegetable, say, over a lifetime period?"[83] In time, this line of argument would evolve into the "toxicity of natural products" defense, as scientists explored the toxicology and pharmacology of organic products.

REREADING *SILENT SPRING*

Jaworski emphasized that scientists knew a great deal about pesticide toxicology and pharmacology and relatively little about natural products: "We really do not know and the plant biochemist knows that there are many naturally occurring organic materials which nobody has ever studied in terms of their pharmacology and toxicology."[84] Within the context of an extended discussion of the importance of a no-effect level as a baseline for toxicological analysis, however, the Monsanto scientist's implication is clear: if natural products proved to be highly toxic (or at least more toxic than some chemical pesticides), should regulators have devoted greater concern to those risks? In other words, regulators should have expanded the study of toxins and no-effect levels to natural products. The well-known biochemist Bruce Ames revisited and refined this line of argument in 1990.[85]

Ribicoff's questions regarding the no-effect level for pesticides like dieldrin scratched the surface of a larger issue in the history of toxicology; namely, the existence of thresholds of effects. The question for regulators was, "Is there a threshold for exposure below which there is no identifiable effect?" Pharmacologists explored a related question: "For a given drug, is there a minimum (measurable) dose below which there is no response?" As we have seen, certain industry representatives accepted the no-effect level as a necessary component for toxicological analysis, whereas others went to great lengths to undermine its significance. The development of technology, in the form of Lovelock's electron capture detector, facilitated finer analysis and the ability to detect exposure in the parts per billion (and eventually, as the ECD was refined, in the parts per quadrillion). But was it possible to identify and measure the effects of these minute exposures?

The gap between toxicology and pharmacology contributed to a sense of frustration among researchers and regulators. In creating the Institute of Toxicology, some toxicologists hoped to close the distance between the two related fields. As Arnold Lehman related, late in 1962 he met with two other toxicologists and by the end of the year they had established a journal and created a society (see above), but the most important step, in Lehman's opinion, was the creation of an Institute of Toxicology on February 1, 1963, at Albany Union Medical College in Albany, New York. Lehman reflected: "The purpose of this institute is to help us in our pesticidal work. There is one very large gap in this

179

REREADING *SILENT SPRING*

research on pesticides. Although our protocol has done very well for the human safety, there is still one gap that we must fill, and that is human pharmacology, the testing of pesticides in humans."[86] For the first time, it would be possible to conduct controlled, pharmacological analyses of pesticides in humans using volunteers at prisons. It is not clear whether Lehman did not accept the validity of Wayland Hayes's research on the effects of pesticides in humans or whether he did not consider the research "pharmacological" (see chapter 2 for analysis of Hayes's studies of the effects of DDT in humans).

Lehman also addressed questions regarding the phenomenon of potentiation or the interaction of two or more insecticides to create a toxicity level greater than the additive effects of exposure. Lehman's contention was that potentiation was an extremely rare occurrence; he noted that scientists had documented only five or six cases since the original discovery. When pressed, Lehman quantified his answer: "It has been known to occur, but it is very infrequent. Now, if we were to take the 125 pesticides and test them against the 2,000 drugs, you would have about a quarter of a million tests to do, thousands of animals and years of time. It would be a tedious test. So far these potentiation effects are always discovered accidentally."[87]

Kenneth DuBois also testified to the committee regarding potentiation. Drawing on data from drug interactions, DuBois defined potentiation as "greater than additive effects of combinations of drugs."[88] He went on to note that no one really considered the possibility that potentiation of the toxicity of pesticides might occur until 1957 when FDA scientists discovered that two insecticides that were permitted in foods caused potentiation when they were present together at certain levels in the diets of animals. On the basis of this finding, the FDA launched a program to evaluate insecticides that were cholinesterase inhibitors in combinations with one another to determine whether potentiation occurred. DuBois recognized that several pairs of insecticides did cause potentiation of toxicity when given in high doses, but he had not identified potentiation of toxicity in experimental animals when pairs of insecticides were added to the diet at the levels that were permitted in various foods. Thus, he concluded, "This indicates that the tolerance levels selected for these pesticides contain a sufficient margin of safety to take

REREADING *SILENT SPRING*

care the unpredictable occurrence of potentiation, at least potentiation caused by a similar type of insecticide."[89]

Of greater concern to DuBois, however, was the possibility that other environmental exposures might contribute to potentiation of toxicity (see chapter 4). He cited the case of radiation, which he and his colleagues at the Tox Lab had studied experimentally. When they exposed animals to low doses of radiation, the radiation stopped the development of enzymes in the liver that would normally detoxify some of the organophosphate insecticides. As a result, the irradiated young animals became hypersensitive to insecticides and also to certain common drugs. DuBois's main thesis was that potentiation could be caused by an unrelated substance, which had much broader implications for environmental health than the discovery that two compounds from the same class (like cholinesterase inhibitors) could cause potentiation.[90] Such implications prompted DuBois to call for a comprehensive research program, on a much larger scale than previous initiatives, dedicated to the study of potentiation to eliminate exposures to any combination of agents that might endanger public health. The scale of the investigation transcended university capacity and the lack of obvious benefits to industry limited corporate investment, and so DuBois recommended that a national research center tackle the problem.

Frawley further clarified the problem of potentiation. He agreed with Lehman that potentiation was a rare occurrence except in cases when doctors administered a number of drugs at pharmacologically active levels. In contrast, pesticides did not appear in the diet at such levels. Indeed, at levels of less than 1/100 of the no-effect level in experimental animals, the potential exposures from food residues would be far below a pharmacologically active level, according to Frawley. He concluded: "I know of no situation again where any potentiation has been demonstrated at several times above the level that pesticides occur in the diet. Potentiation can, however, be demonstrated at high levels approaching the acute toxic dose, or even maybe a tenth of the acute toxic dose. The potentiation indeed occurs, but that is a pharmacologically active dose."[91] Once again, toxicity hinged on the dosage or exposure, leaving the question of potentiation as a result of exposures to insecticides at the levels of food residues unresolved.

181

REREADING *SILENT SPRING*

Before and certainly after the publication of the provocatively titled *100,000,000 Guinea Pigs,* there was a pervasive notion that pesticides and other toxics arrived at the marketplace with minimal testing for toxicity. In *Silent Spring,* Carson lent a certain amount of credence to the view that industry and governmental agencies like the USDA were mindfully exposing countless animals and numerous agricultural workers and their families to understudied chemicals across America. Corporations struggled to represent their testing efforts before the public release of a new chemical insecticide. Industry representatives detailed the process of toxicity testing for new pesticides along with the approximate time and cost of such analysis. Frawley noted that the total approximate cost for the development of a new insecticide was two million dollars and the process required approximately five years from initial test tube synthesis to government approval. Other industry representatives were more specific in presenting the steps that led to the development of a new pesticide. For example, Johnson (Dow) presented a detailed chronology of the development of a pesticide. The average time from laboratory synthesis (in the test tube) to the first date of sale typically took six to seven years. Over the course of ten years, Dow Chemical tested an average of 4,150 compounds per year. To capture the range of biological activity, the compounds were tested on at least 50 different living organisms. Out of more than 4,000 chemicals, a yearly average of 27 passed to stage 2 testing. In stage 2, the studies of metabolism began with compound radioisotope labeling. An average of 2 chemicals (of an original 4,150) passed from stage 2 to stage 3. The cost of testing up to stage 3 amounted to $500,000, approximately. Stage 3 launched the two-year dietary feeding toxicity tests in dogs and rats. It is at this point that Dow Chemical obtained symptomology and treatment information for doctors' review to protect Dow researchers and later customers. Johnson specified the broad ramifications of this research: "We consider the potential crop application hazards, the drift, the persistence in soil, and leaching to ground waters."[92] Other industry scientists confirmed that they too evaluated these aspects of environmental toxicology. In stage 4, researchers at Dow Chemical worked directly with agricultural experiment stations and the USDA as well as consulting laboratories for specific toxicological investigations. Research on chronic toxicity, analytical methods, and residues (using radiochemical methods) continued through

REREADING *SILENT SPRING*

this stage. Moreover, three-generation reproduction studies commenced. Johnson noted that Dow Chemical had been conducting these studies for the previous three to five years. At least in part as a response to *Silent Spring*, the PSAC had called for incorporation of two-generation reproduction studies. Monsanto's Jaworski initially bristled at the imposition of three-generation studies but demurred when informed that the PSAC called for two-generation studies, which Monsanto and other companies conducted.

To determine potential for environmental impact, Dow scientists examined effects on snails, fish, *Daphnia,* and algae, beginning about three years ahead of the projected release. Johnson made explicit reference to food chains: "We hope to help assess the hazards of stream and lake pollution and get this information on effects on biological life chains."[93] The first-year progress report on the two-year toxicology followed with a real chance that 1 of the 2 chemicals that reached stage 4 would not reach market, or, in other words, 1 new insecticide, out of an average of 4,150 per year, actually survived to reach the market. Crops from at least ten experimental studies across the country underwent analysis, and stage 4 also included meat and milk residue studies, if appropriate, as well as human skin sensitization studies and inhalation studies, if mists or dusts were to be employed. After they wrote the label, the Dow scientists prepared the petition to go to the FDA: "In this report are included the performance information, the proposed use and labeling, the dietary feeding toxicology report, the wildlife report, the residue report, the analytical report, any information on symptomology and treatment which should go to a physician, reproduction studies, and toxicology report."[94] The final steps of stage 4 involved education of the salespeople, the resellers, and the applicators. Stage 5 involved shipment, sales, and the launch itself. Johnson concluded his comprehensive review of the development of new pesticide with an overview of the process: "Now, this represents an effort of about 5 to 7 years—if we are lucky it is 5 years—and perhaps between $2 and $3 million if food crops are involved."[95]

Jaworski was more succinct in his exposition regarding new pesticide development. After noting that Monsanto was among the first chemical companies to manufacture DDT (a practice later abandoned), he acknowledged that his company was one of the largest manufacturers of

183

herbicides, insecticides, and animal feed additives in the United States as well as one of the five major manufacturers of parathion in the world. Out of 7,000 to 10,000 chemicals screened each year, 50 to 100 received advanced laboratory evaluation and no more than 2 or 3 survived the first year of field study, after which they would be studied in greater detail for efficacy, toxicology, and residues. The entire research department participated in the extensive analysis of chemicals over the course of four to seven years to commercialize a new pesticide at a total cost of between 1.5 and 2.5 million dollars. Jaworski concluded: "We cannot afford to develop poor performing or hazardous materials nor can we afford second-rate research if we are to maintain good relations with the ultimate consumer of our products. Safety, product performance, and confidence by the consumer in our industry demand this high level of professional endeavor."[96]

Each of the research scientists from the chemical corporations indicated the considerable scope (in terms of time and cost) of research and development of a new pesticide. The process involved thousands of chemicals, four to seven years of laboratory and field analysis, and from one to three million dollars. If during the course of investigation, scientists identified a problem with an insecticide, development could be abandoned. Johnson cited several examples of promising potential insecticides that Dow removed from the later stages of development as a result of toxicological analysis (see above). Nevertheless, Senator Ribicoff remained skeptical, noting a report from *Chemical Week Reports* (June 1, 1963) that predicted that pesticides could be a two-billion-dollar market by 1975. He argued, "A $2 billion industry, it would seem, has quite a bit of responsibility to make a thorough study of its products and also look at the side effects or no effects."[97] Ribicoff's statement placed the claims of industry scientists regarding the costs of pesticide development in broader financial context.

Although five to seven years and two to three million dollars devoted to research and development sounded impressive, the question remained whether such research was sufficient to insure the safety of consumers of chemical insecticides as well as exposed wildlife. The answer to this question arrived through the prolonged process of litigation (see chapter 7).

As Daniel, Langston, Oreskes and Conway, and Rosner and Markowitz and others have brilliantly demonstrated through numerous inci-

REREADING *SILENT SPRING*

sive examples, industry successfully captured regulatory agencies in the U.S. across the twentieth century to the detriment of the health and wellness of millions of Americans and others worldwide.[98] In the case of pesticides, Daniel argued: "[USDA's Agricultural Research Service (ARS)] possessed enormous power, for its label approval function licensed pesticide formulations. It garnered enormous power in its multiple roles as clearinghouse, coordinator, regulator, and research center. To have their way, ARS bureaucrats bullied, plotted, lied, and misled. A culture emerged within the service that justified pesticides at all costs, and staffers bent research, reports, and testimony to serve this mission."[99] As a result, Daniel condemned the ARS as follows: "The ARS was assigned to umpire the debate and assure public protection, yet it had become simply the handmaiden of the chemical companies."[100]

Scientific uncertainty played a critical role in the strategies of various industries to evade or delay regulation. By emphasizing uncertainty in toxicological effects other than acute, companies could stymie regulators. But these authors have shown in case after case that regulatory agencies like the USDA and the FDA were "captured" as their designated representatives adopted views entirely compatible with those of industry. Exceptions to this rule—Kelsey at the FDA, Hueper at the NCI, and Clarence Cottam at the FWS, as well as Carson—stand in sharp distinction to a pervasive trend.

As convincing as I find these case studies, my argument is simpler. When it came to organophosphates, no one seriously argued that they were safer than chlorinated hydrocarbons like DDT—and in that group I include toxicologists, regulators, industry representatives, and environmental advocates. In fact, as I have shown, in statement after statement, nearly everyone who testified in the various pesticides hearings readily acknowledged that, with the exception of malathion, organophosphates posed greater risks to humans and wildlife. For a pesticide like parathion, scientists, regulators, and industry representatives all agreed that it posed risks at the minutest exposures levels, at one or two parts per million! And yet, as most experts also noted, organophosphates had one notable advantage over DDT and the chlorinated hydrocarbons: organophosphates broke down into relatively harmless components over the course of weeks or even days, whereas chlorinated hydrocarbons accumulated in ecosystems and the bodies of wildlife and humans.

185

REREADING *SILENT SPRING*

In retrospect, Geiling's advice to be mindful of an obligation to the public seems prescient as debates surrounding pesticides moved from the realm of science and policy to public discourse. Geiling offered his words of wisdom at a time when toxicologists were establishing an independent professional identity. With the publication of *Silent Spring* and subsequent publicity, including a nationally aired news program, the public discovered the risks of DDT and other synthetic insecticides as well as the science of toxicology. Rachel Carson's unassuming yet forceful prose revealed hazards attendant with the indiscriminate use of DDT and other chlorinated hydrocarbons, but some of her most disturbing case studies involve organophosphates. Her case for curbing pesticide use applied to both chlorinated hydrocarbons *and* organophosphates. *Silent Spring* and the public outcry that followed inspired further study at the federal level, first by the PSAC and then the congressional Committee on Interagency Coordination. As in other hearings, few questioned the considerable benefits of pesticides, and most witnesses couched evaluations of risks in light of benefits, but most witnesses, including Carson herself, acknowledged the considerable dangers associated with organophosphates. A familiar cast of scientists from the USDA, the FDA, and the Tox Lab presented their findings regarding organophosphates, the no-effect level, and potentiation (which was most common among organophosphates). In an attempt to defend their safety record, several representatives of the chemical industry presented the multistage and multiyear process, involving literally thousands of chemicals, through which a company identified, tested, and marketed a new insecticide. With a sharper picture of toxicological risk presented in layman's terms in *Silent Spring* and thoroughly analyzed by the PSAC and Congress, the pathway to further regulation appeared clear.

CHAPTER 7

Pesticides and Toxicology
after the DDT Ban

In *Silent Spring,* Rachel Carson meticulously described a typology of chemical insecticides. Most commonly used insecticides fell into one of two groups: chlorinated hydrocarbons (or organochlorines) like DDT and dieldrin and organophosphates like malathion and parathion. Both groups of chemicals had implications for the health of humans, wildlife, and ecosystems. Carson revealed how DDT and other organochlorines bioaccumulated in the environment and biomagnified in organic systems, until reaching toxic levels in topline predators such as ospreys, brown pelicans, and of course bald eagles. Organophosphates did not typically bioaccumulate, but exposure to the potent nerve agents resulted in cholinesterase inhibition by disrupting the normal function of an enzyme that was critical to normal nerve function. Organophosphates could cause tremors, convulsions, and even death in wildlife and humans. In case after case, Carson cited organophosphate poisonings of farm workers and wildlife. As we saw in the previous chapter, numerous witnesses from a diversity of perspectives testified to the extreme toxicity of organophosphates before the PSAC and in congressional hearings.

Nevertheless, Americans in general, and legislators in particular, took a different message from *Silent Spring,* namely, that DDT was the most harmful insecticide for its effects on wildlife, particularly birds. Moreover, Americans and American legislators became concerned about environmental cancer. It would be naïve to suggest that *Silent Spring* was

187

PESTICIDES AFTER THE DDT BAN

the only factor in raising concern regarding environmental cancer. Congressional hearings devoted to possible risks, including cancer associated with pesticides beginning in 1951, resulted in passage of the Miller Amendment and the Delaney Clause (see chapter 5). But *Silent Spring* certainly crystallized this concern in the minds of many Americans, along with worries about environmental contamination and wildlife effects. Against such a backdrop, the banning of DDT late in 1972 concluded one of the greatest success stories of the environmental movement. But if history teaches us anything, it is to be wary of simple stories.

According to historian Thomas Dunlap, *Silent Spring* (and the public controversy in its aftermath) failed to effect significant changes in pesticide use and regulation in the short term.[1] The USDA did curb its extensive spraying programs, but other government agencies and farmers continued to employ DDT. In the cases where DDT was discontinued, it was replaced with more toxic chemicals like the organophosphates. During the remainder of the 1960s, environmentalists went to court several times in an attempt to control or stop the use of DDT. The recently established Environmental Defense Fund (EDF) pursued litigation against various uses of chlorinated hydrocarbons on several fronts. In Michigan, the EDF entered litigation against spraying programs that utilized DDT and dieldrin against Dutch elm disease. EDF's action against dieldrin was thrown out of the court of appeals by the judge, but by taking the case to the Michigan Supreme Court, EDF was able to delay the spraying campaign until the best time for spraying had passed. The case against DDT in Dutch elm disease control did not reach the courts, because each of fifty-six communities agreed to stop applying DDT. EDF next took up the fight against dieldrin in Wisconsin, where the courts were marginally more interested than those in Michigan. By the end of preliminary discussions, however, a judge had rejected the arguments of EDF's lawyers, cut their presentation short, and allowed for dieldrin spraying to continue. Near the end of 1968, Wisconsin did allow an extended hearing on whether DDT could contaminate water. Dunlap has argued that the six months of this hearing saw a sea change in Americans' trust in government and their willingness to support groups that questioned government policies. By the end of 1968, there was a groundswell of support for the EDF's efforts against DDT and other chlorinated hydrocarbons.[2]

188

PESTICIDES AFTER THE DDT BAN

Neither the passage of the National Environmental Protection Act (1970) nor the establishment of the Environmental Protection Agency (EPA) in October 1970 ended the battle against DDT. The EPA inherited USDA's authority and staff in pesticide regulation, and the newly formed federal agency became a target for litigation. EDF returned to court, bringing suit against the EPA in *EDF v. Ruckleshaus* (D.C. Circuit 1971), in which EDF sought review of the failure to cancel the registration of DDT and to stop its use during cancellation hearings. During the course of this case, EDF strengthened its argument sufficiently so that the EPA had to take significant action against DDT. The judge and two colleagues ordered William Ruckleshaus (EPA administrator) to end all uses of DDT immediately. Although complying initially, Ruckleshaus refused to suspend registrations after a sixty-day review. Instead, at the behest of the U.S. Court of Appeals, the EPA would hold hearings to determine whether DDT posed a threat to human health. After the initial hearings, Ruckelshaus acknowledged that evidence that DDT was carcinogenic in lab animals did not prove that DDT was threat to humans. Moreover, he believed that "a quick and total ban on DDT would force farmers to resort to highly toxic alternatives."[3] Nevertheless, in June 1972, after yet another lengthy hearing, Ruckleshaus banned the remaining uses of DDT on crops. He did, however, allow it to be used in cases of urgent public health, such as emergency quarantine. He also allowed for it to be manufactured for export. At the same time, EPA suspended most uses of dieldrin, but it took more than two years before the agency announced a ban on the manufacture of dieldrin and aldrin.[4] More than a decade had passed since Rachel Carson alerted Americans to the environmental and health risks of synthetic insecticides. Most, if not all, of the legislative effort during the years leading up to the ban on DDT, dieldrin, and aldrin was concentrated on the persistent chlorinated hydrocarbons. The extensive toxicological research on organophosphates, Carson's significant concern regarding wildlife and human effects, and extensive testimony in hearings at the federal level would naturally lead one to expect comparable legislative scrutiny for these highly toxic chemicals. But such examination was delayed.

On October 14, 1964, in a symposium on environmental health hazards at the 148th meeting of the American Chemical Society, DuBois, who had linked the organophosphates to cholinesterase inhibition, pre-

189

sented his theory that humans probably developed a resistance to toxicity from the prolonged low consumption of organophosphate insecticides. DuBois reasoned that since dogs and rats developed resistance to organophosphates after just ten days of exposure, humans exposed regularly to such chemicals might also develop resistance. Experimentally, DuBois found that when the acetylcholine level was increased over a period of several days, cells that it would normally have stimulated became somewhat resistant to its effects. He concluded: "This resistance explains why persons who are occupationally exposed repeatedly to organophosphates often fail to show symptoms of poisoning although their ability to detoxify acetylcholine is impaired."[5] In time, DuBois's view would change.

Five years later, DuBois warned that pesticide poisonings were widespread in the United States and more common in urban and suburban areas where people valued highly manicured lawns of grass. He argued that thrift and federal regulations compelled farmers to minimize pesticide use. According to DuBois, a host of mild symptoms, including blurred vision, sweating, excessively watery eyes, excessive salivation, diarrhea, and tightness in the chest, could all be signs of pesticide poisoning (such a statement resonated with Kallet and Schlink's worries about chronic lead poisoning thirty-five years earlier and with Morton Biskind's concerns regarding DDT). He repeated earlier warnings regarding the effect of pesticides on drug efficacy, and he added weight loss to the list of risk factors (losing weight could release pesticides from stored fat into the bloodstream). DuBois identified chlorinated hydrocarbons as the cause of these problems, but he mentioned the significant risk of direct exposure to organophosphates.[6]

After the EPA banned DDT from use in the home and on lawns, trees, and gardens near the end of 1971, DuBois warned that substitutes like parathion might increase the incidence of poisonings. DuBois believed that the ban reflected a shift in priorities from medical threats to man to concern about chemicals in the environment. He worried that physicians would become increasingly involved in the treatment of pesticide poisonings in urban communities.[7] He was concerned mainly that parathion or some other highly toxic organophosphate would be substituted for DDT, and he noted a striking increase in the number of cases of poisoning by such chemicals. With respect to human toxicity alone,

DuBois believed that DDT was the safer choice. He noted that there had been a "remarkably small" number of cases of acute human DDT poisoning, and that the most significant risk was mild (and reversible) liver damage.[8]

Like many other scientists, DuBois struggled to reconcile data from wildlife studies with his own research on human health effects. His research on cholinesterase inhibition and organophosphate insecticides indicated that parathion and other chemicals in the class were highly toxic to humans. Replacing DDT (of minimal acute mammalian toxicity and uncertain chronic toxicity in humans) with a known hazard like parathion made no sense, regardless of mounting evidence of chronic effects of DDT in wildlife. The manufacturers of DDT presented a similar argument in defending their product. They, too, reasoned that DDT would have to be replaced by highly toxic chemicals like parathion, thus increasing risks to consumers. Ruckleshaus noted the hazards of parathion and other organophosphates: "The introduction into use of organophosphates has, in the past, caused deaths among users. . . . A survey conducted after the organophosphates began to replace chlorinated hydrocarbons in Texas suggests a significantly increased incidence of poisonings."[9] Yet DuBois's warnings went unheeded. His death from lung cancer in 1973 at the relatively young age of fifty-five effectively ended the rich history of the Tox Lab at the University of Chicago, but its legacy continued at numerous universities across North America through the efforts of the many toxicologists who had received their education and training at Chicago. Legislation focused on the environmental effects of pesticides despite DuBois's concerns.

Amid continuing debates regarding DDT, several prominent court cases regarding pesticides, and nearly two years of hearings focused on deficiencies with FIFRA, Congress passed the Federal Environmental Pesticide Control Act (FEPCA) of 1972.[10] The product of political compromise on the part of the Nixon administration, the bill that led to FEPCA was criticized by both farmers and environmentalists.[11] As we have seen, FIFRA required registration once a pesticide was determined to be effective and safe when used as directed. John Wargo noted that FEPCA, in contrast with FIFRA, required manufacturers to "demonstrate that a pesticide would perform its 'intended function' and, 'when used in accordance with widespread and commonly accepted practice,' would not cause 'unreasonable adverse effects on the environment.'"[12] "Unreasonable adverse effects on the environment" applied to both man and the environment and took into account the economic, social, and environmental costs and benefits of the use of any pesticide. Since Congress neglected to define key terms such as "risk," "cost," and "benefit," FEPCA left the EPA in a precarious regulatory position.

The 1972 law did establish two categories for pesticides: general use and restricted use. General use pesticides were unlikely to cause adverse effects; restricted use applied to the more dangerous pesticides that could cause "unreasonable adverse effects" even if used in accordance with label instructions. According to FEPCA, restricted use pesticides could be used only "under direct supervision of a certified applicator." However, the law allowed for delegation to any competent person under the supervision of a certified applicator. Wargo commented: "The health and safety of the public thus rests on whether the applicator, trained or not, complies with the special precautionary measures provided with the restricted compound."[13] Agricultural chemical manufacturers supported the new legislation because they approved of the certified applicator provision, which was applied to a limited number of chemicals. Designating some pesticides as restricted removed regulatory pressure from the majority of other chemicals. A 1975 amendment further relaxed the restrictions by forbidding the EPA from testing "an applicator's knowledge of a pesticide's potential to injure health or the environment."[14]

FIFRA provided the prerogative to suspend or cancel the registration for a pesticide if it posed an "imminent hazard to the public," without further clarification as to the precise meaning of this phrase. FEPCA, however, defined "imminent hazard" as "a situation which exists when the continued use of a pesticide during the time required for cancellation proceedings would be likely to result in unreasonable adverse effects on the environment or will involve unreasonable hazard to the survival of a species declared endangered by the Secretary of the Interior."[15] With the passage of FEPCA, it became the EPA's responsibility to reregister nearly sixty thousand chemicals. When the EPA identified that a pesticide had an adverse effect on the environment, it sent cancellation notices to the manufacturers, but such notices merely signaled that the EPA had initiated a review of all available data regarding the chemical's

PESTICIDES AFTER THE DDT BAN

toxicological and ecological effects. Given the poor quality of data collected by the USDA under FIFRA, such reviews could extend for years, during which time manufacturers continued to produce and distribute the chemical.

Even after it suspended or canceled a compound, the EPA typically permitted the use of the existing supplies without recall. FIFRA included an indemnification clause that required EPA "to purchase, collect, and dispose of any recalled products." Given that most pesticides had been registered by the USDA under the provisions of FIFRA, it proved to be a challenge for the EPA to obtain high-quality evidence regarding toxicity and fate of pesticides for re-registration decisions as well as decisions to classify pesticides for general or restricted use. Most of the data that the USDA collected under FIFRA addressed the efficacy and efficiency of chemicals in controlling pests rather than toxicity and environmental fate. Moreover, much of the data was more than a decade old. Wargo noted, "EPA's role has therefore been to manage the scientific process, judge the potential for unreasonable hazard, and then adjust existing use entitlements."[16]

As noted above, FEPCA was supported by industry, especially since one stipulation involved indemnification, the compensation of manufacturers for stocks of pesticides that were either canceled or suspended. Chemical companies viewed indemnification as a form of insurance that was necessary for them to invest in the research and development of new pesticides, a potentially costly enterprise. Some environmentalists came to look upon indemnification as a deterrent to federal suspension or cancellation, since the Office of Pesticide Programs within EPA would bear the costs of indemnification. In fact, as Wargo has shown, between 1972 and 1988, EPA indemnified manufacturers of several herbicides and insecticides, spending in excess of sixty million dollars in the process.[17] In 1988, Congress revised the indemnification clause and ended financial protection for manufacturers yet continued to protect farmers and applicators who could be indemnified by the regular federal Judgment Fund as directed by the Treasury Department rather than the EPA Office of Pesticide Programs. Despite the challenges posed by classification of pesticides for general or restricted use, re-registration, and indemnification, FEPCA guided pesticide regulation from 1972 forward. Significantly,

193

PESTICIDES AFTER THE DDT BAN

FEPCA established, as part of federal statute, the protection of the environment as well as human health from the damaging effects of pesticides.[18]

The EPA was forced to apply FEPCA shortly after it was passed by Congress. The EDF initiated the legal action in 1968 (before FEPCA) by petitioning the FDA to reduce to zero all food tolerances for two closely related chlorinated hydrocarbons—aldrin and dieldrin—based on evidence from the Shell Chemical Company study indicating that the compounds caused cancer in mice. The day after EPA opened for business, EDF requested it to suspend and cancel all uses of aldrin and dieldrin. Ruckelshaus, the EPA administrator, complied by issuing a notice of intent to cancel all registrations of the two pesticides. The companies that held the registrations for aldrin and dieldrin demanded review by the National Academy of Science (NAS), which they could do under FIFRA. After nearly two years, the NAS report reached the conclusion that the pesticides appeared to pose no threat to human health even in their uses as a corn insecticide. EDF next requested that the EPA suspend uses of aldrin and dieldrin through the D.C. Circuit Court in 1972. Cancellation hearings for the two pesticides began in August 1973 after the passage of FEPCA. The new EPA administer, Russell Train, reviewed testimony generated over the course of twelve months (more than 35,000 pages). Train concluded that aldrin and dieldrin represented an "imminent hazard to man and the environment." After an expedited suspension hearing, a judge recommended suspension of all uses. Train agreed to the proposal but Shell Chemical Company appealed, and in April 1975 the D.C. Circuit Court upheld the ruling.[19] The cancellation of aldrin and dieldrin took place between 1968 and 1975, which meant that the process began under the rules of FIFRA and concluded under FEPCA.

As with the cancellation of DDT, cancelling aldrin and dieldrin proved to be a labored and tortuous process of requests to FDA then EPA, chemical company appeals, proposed rulings, additional appeals, and, eventually, resolution. As of 1975, the EPA had canceled registrations on several of the persistent chlorinated hydrocarbons that posed the most significant threats to the environment. Meanwhile, many other pesticides remained on the market, including the highly toxic organophosphates. And chemists strove to develop synthetic forms of naturally occurring insecticides.

194

PESTICIDES AFTER THE DDT BAN

Banning DDT and other chlorinated hydrocarbons along with some of the organophosphates in the early 1970s left economic entomologists, farmers, and public health officials with a significantly reduced palate of chemical insect control options. Recall from chapter 1 the discussion of insecticide options that preceded the development and proliferation of DDT and other synthetic insecticides. Before there was DDT, before there were heavy metal insecticides such as lead arsenate, there was pyrethrum. Pyrethrum was a potent naturally occurring insecticide, but production at the levels required to control insects on crops proved problematic. By the 1940s, chemists were committed to developing synthetic analogs to pyrethrum, the first of which—allethrin—saw the light of day in 1949. Along with its alcoholic component and other forms, allethrin served as an important alternative to the natural form of pyrethrum, when it was not available.[20] But as we have seen, DDT, other chlorinated hydrocarbons, and the organophosphates surged onto the market and quickly dominated agricultural and public health insect control campaigns in the late 1940s and 1950s, leaving very limited utility for the first synthetic pyrethroid. John E. Casida, a chemist at the University of California–Berkeley, captured alletrin's significant disadvantage to DDT to American consumers thusly: "Allethrin, the first synthetic pyrethroid, was useful for household pests if you accepted a many-step synthesis, whereas DDT controlled almost every pest and could be made in one or two steps at only a small fraction of the cost. No wonder billions of kilograms of chlorinated hydrocarbons including DDT were used!"[21]

In 1966 the synthesis of pyrethroid-like compounds took a significant leap forward through the efforts of Michael Elliott and a group of organic chemists at the Rothamsted Experimental Station in Harpenden, England. Elliott successfully produced the first compounds with properties superior to the properties of the natural ester, indicating potential as practical insecticides. The most valuable property of bioresmethrin was that it combined great insecticidal activity with very low mammalian toxicity, which led to the commercial production of resmethrin and bioresmethrin.

In 1970 Elliott presented his findings regarding the considerable potential of new synthetic pyrethroids at the International Conference on Alternative Insecticides for Vector Control, sponsored jointly by Emory

195

PESTICIDES AFTER THE DDT BAN

Table 2

Toxicity of Parathion and Bioresmethrin to Housefly and Rat

	LD_{50} (mg/kg) to	
	Housefly (topical application)	Rat (oral administration)
Parathion	1	5
Bioresmethrin	0.25	8,000

Source: Michael Elliott, "The Relationship between the Structure and the Activity of Pyrethroids," *Bulletin of the World Health Organization* 44 (1970): 315.

University, CDC, and WHO, a conference that focused predominantly on so-called anticholinesterase insecticides, which is to say the organophosphates and carbamates, although pyrethroids and chlorinated hydrocarbons also received attention. To make the case for the potential of synthetic pyrethroids, Elliott compared the toxicity of bioresmethrin and parathion (still very much in use in 1970) (see table 2). Elliott interpreted the table noting that bioresmethrin was more toxic than parathion to houseflies but much less toxic to rats. To interpolate what must have been a stunning revelation to both economic entomologists and toxicologists, bioresmethrin was four times more toxic to houseflies (the target organism) than parathion and less toxic to rats by more than three orders of magnitude! Parathion was widely considered to be one of the most toxic chemicals known to man, and its toxicity applied equally well to mammals and birds as to insects. As an aside, Elliott noted that synthetic pyrethroids were "non-persistent," which distinguished them from chlorinated hydrocarbons. Economic entomologists, however, viewed the blessing of rapid decomposition as a curse. Allethrin, bioallethrin, resmethrin, bioresmethrin, and others were all unstable in air and light, a property that significantly limited the utility of these compounds in the field, particularly against pests of agricultural crops, despite the manifest advantages of these chemicals (potency against many insect species, rapid action, and low mammalian toxicity). Elliott and the researchers at the Rothamsted Experimental Station endeavored to synthesize esters that were more stable in light by a factor ten to one

PESTICIDES AFTER THE DDT BAN

hundred than previous pyrethroids, while retaining the strong activity against insects and the low toxicity to mammals.[22]

In 1972, Elliott and his colleagues discovered an "exceptionally valuable combination" of esters and alcoholic components. They called the new compound permethrin. It was more active against many insects than had been predicted from its components. Moreover, permethrin was more stable in air and light than other potent pyrethroids, and it had lower mammalian toxicity than other esters created from the same acid. Elliott would later conclude that permethrin was "stable enough to control insects in the field as efficiently as established organophosphates, carbamates, and organochlorines, many of which it surpassed in potency."[23]

Elliott compared permethrin with the most significant chlorinated hydrocarbons, organophosphates, and carbamates that were the most popular insecticides. At 0.7 µg/g, The LD_{50} of permethrin to insects was lower than that for any other insecticide. Parathion showed the next lowest LD_{50} at 1 µg/g for insects. Parathion's LD_{50} for rats was 11 mg/kg! In contrast, the LD_{50} for rats for permethrin exceeded 1,000 mg/kg. Of the dozen or so insecticides that Elliott included in the table, only malathion had an LD_{50} for rats that was greater (1,400 mg/kg). Nevertheless, permethrin stood apart from all the other insecticides on the basis of yet another metric: the ratio of LD_{50} of rats to insects, which provided an index to an insecticide's relative effectiveness and safety.[24]

With such high toxicity to insects and such low toxicity to mammals, permethrin and other synthetic pyrethroids had great potential as agricultural insecticides. Laboratory and field tests indicated that permethrin could effectively control insects of various orders, including moths, mosquitoes, flies, and ants. Beyond its strong potential for uses in agriculture, permethrin showed a wide range of activity against veterinary parasites. Researchers found that the new insecticide killed 100 percent of cockroaches (*Blatella germanica*) over the course of twelve months, when applied to plywood in the amount of 300 mg/m². Permethrin also showed promise in the control of dairy barn pests, as reflected in this statement: "Preliminary results indicate that in hand spray applications to the entire body surface of lactating dairy animals, at a level needed for adequate fly control, residues of permethrin in milk are unlikely to pose a problem."[25] Recall that the USDA officially discouraged comparable uses of DDT as early as 1949. As with other insecticides, and

PESTICIDES AFTER THE DDT BAN

particularly in cases in which other insecticides had been used intensively, insects developed resistance to pyrethroids. After recommending caution with regard to the development of resistance as a result of overenthusiastic application, Elliott optimistically concluded his review of the toxicity and potential for pyrethroids: "In favorable circumstances however, synthetic pyrethroids should help to control more insect pests in the future with smaller hazard to man and the environment than earlier, widely used pesticides."[26]

The promise of synthetic pyrethroids continued to emerge as the insecticides were developed for agricultural and household use during the 1970s. Toxicological profiles for the chemicals still appeared to be quite favorable, especially in comparison with chlorinated hydrocarbons and organophosphates. Initial assessments of mammalian toxicity were borne out by subsequent studies. Avian toxicity also appeared to be quite low, but toxicity for fish was strikingly high (sensitivity ranged into the parts per billion). Toxicologists recommended caution to avoid the contamination of lakes and streams as well as commercial fisheries. Besides fish, honeybees were also highly susceptible nontarget organisms. Beneficial aquatic insects and crustaceans were also potentially vulnerable to pyrethroids.

As with other organic insecticides, resistance presented a potential problem. Neither organophosphates nor carbamates seemed to introduce cross-resistance to pyrethroids, but houseflies on Danish farms exhibited considerable cross-resistance to pyrethroids after developing resistance to DDT. Cross-resistance among pyrethroids was potentially serious. Selection with bioresmethrin of a pyrethroid-resistant field strain resulted in a resistance factor of 1,400-fold. This same strain revealed a resistance factor of 60,000-fold to decamethrin, despite no previous exposure.[27]

In 1978, Michael Elliott reviewed the status of insecticide development. He argued that conventional insecticides were necessary to protect food supplies and other agricultural crops and to control disease. The only two natural insecticides that had significant histories of use were pyrethrum and nicotine, derived from the leaves of plants in the nightshade family (Solanaceae). Elliott noted that nicotine was still used in surprisingly large quantities. His surprise may have followed from the insecticide's variable toxicity to insects. Nicotine was much more toxic

PESTICIDES AFTER THE DDT BAN

to silkworms than houseflies, for example. More likely, the continued use of nicotine surprised Elliott because it was "one of the most rapidly acting and deadly poisons known against man."[28] Despite efforts to find chemical analogs to nicotine, none were less toxic to humans. Nevertheless, agricultural chemists continued their efforts to develop synthetic nicotine analogs. In contrast, rotenone, another naturally occurring pesticide extracted from the roots and stems of several tropical plants, had a low toxicity to mammals, but like nicotine it was far more toxic to certain insects, for example the mustard beetle, than to the housefly or the honeybee. And, like nicotine, no useful chemical analogs had been synthesized. Mammein (extracted from the seeds of *Mammea americana*), was another natural insecticide, but no synthetic analogs emerged. Elliott reviewed and dismissed other natural insecticides before turning to the synthetic insecticides.[29]

After briefly reviewing the history of the introduction of DDT and the chlorinated hydrocarbons, as well as of the organophosphates and the carbamates, Elliott noted that no other major groups of insecticides had been developed. Then he turned to a discussion of the synthetic pyrethroids and how he and his colleagues at Rothamsted had developed the new class of insecticides. As in his other papers, Elliott noted that the synthetic pyrethroids represented some of the most toxic insecticides, while presenting remarkably little risk to mammals. To illustrate this point, he listed a range of insecticides and other poisons in order of median effective doses to mammals and insects. Like other, similar charts, this one supported Elliott's fundamental argument, but he extended its implication. "Pyrethroids therefore constitute a broad class of lipophilic insecticides which promise to complement the more polar organophosphates and carbamates and to replace the organochlorine compounds for certain applications."[30]

To underscore the low toxicity of pyrethroids to mammals, Elliott expanded the chart, presented earlier in the paper, that showed toxicity to insects and mammals. As in other analyses, pyrethroids appeared among the least toxic insecticides to mammals and among the most toxic to insects. Recommended rates of application for pyrethroids were also much lower than rates for organochlorines, organophosphates, and carbamates. The low rates of application meant that pyrethroids would produce less environmental contamination than organochlorines even if

199

PESTICIDES AFTER THE DDT BAN

pyrethroids shared persistence with the organochlorines. But the nonvolatile pyrethroids broke down rapidly in soil, which reduced the likelihood that appreciable residues would accumulate.[31] By way of conclusion, Elliott emphasized the key similarity between the organophosphates, the carbamates, and the pyrethroids: "The greatest number of practical compounds has been developed in the classes where the groups attached to a central function can be varied widely, to give a range of diverse, but related compounds. The organophosphates, carbamates and pyrethroids are all of this type. The more active and stable pyrethroids have extended the range of useful insecticides."[32] Despite such optimistic claims regarding the synthetic pyrethroids, the ban of chlorinated hydrocarbons like DDT, dieldrin, and aldrin left farmers with limited viable options for pest control.

In *Silent Spring*, Carson recommended judicious use of insecticides and the exploration of alternative methods, such as biological control. To what extent did farmers heed Carson's recommendation? How did regulatory intervention shape use patterns in the aftermath of the ban on DDT and organochlorines? Such questions return us to one of the prominent themes of this book: namely, that throughout the twentieth century, Americans demanded insect control to boost agricultural productivity and protect against insect-borne disease. It is possible to track insecticide use in the United States at five-year intervals between 1966 and 2002, and patterns in insecticide use can provide partial answers to these questions.

In 1966, twelve of the most heavily used insecticides (by weight of active ingredients only) were organochlorines. Farmers applied more than 75 million pounds of toxaphene, DDT, and aldrin, which were the three most popular insecticides. Most of the nearly 35 million pounds of toxaphene was used on cotton. Farmers also deployed nearly 30 million pounds of the five most popular cholinesterase inhibitors: a carbamate (carbaryl) and four organophosphate insecticides (ethyl parathion, methyl parathion, diazinon, and malathion). Some farmers used paraffinic oil as an alternative to other chemical insecticides. Oil killed insects by smothering them in place. As we have seen, ethyl parathion or parathion was one of the most toxic chemicals known, and the toxicity of methyl parathion is similar. Farmers applied more than 8 million pounds of both of these insecticides. The toxicity of the carbamate insecticide,

200

carbaryl (aka Sevin), for humans and other animals, including birds, was very low, even nontoxic, but nontarget insects were highly susceptible. In using 12 million pounds of carbaryl, farmers may have shielded farm workers and wildlife. Diazinon had a classification of toxicity class II (moderately toxic) or toxicity class III (slightly toxic), but it was highly toxic to birds, fish, aquatic organisms, and nontarget insects, notably bees. Unlike most organophosphate insecticides, malathion posed slight toxicity to humans as well as to wildlife. Despite Rachel Carson's warning in *Silent Spring* and the hearings following the recommendations of the PSAC, the use of pesticides rose to new levels in 1966. Chlorinated hydrocarbons appeared to swamp other insecticides, but the millions of pounds of highly toxic organophosphate insecticides posed significant risks to farm workers and wildlife.

By 1971, with the wheels of regulation grinding slowly, organochlorine use had begun to fall. DDT use in agriculture, for example, had dropped to 14.3 million pounds, largely as a result of declining use in cotton growing, a decrease that was partially offset by increases in other insecticides.[33] Similarly, aldrin use had fallen to less than 8 million pounds. Toxaphene use, however, had risen to a new high of 37.5 million pounds. Thus farmers used just under 60 million pounds of these insecticides, down from 75 million five years prior. Meanwhile, the use of organophosphate insecticides had climbed. Methyl parathion led this group with 27.5 million pounds (more than triple the amount applied in 1966). Use of the considerably less toxic carbaryl had also risen to nearly 18 million pounds, but use of ethyl parathion had climbed more than a million pounds to about 9.5 million pounds. With the application of more than 4 million pounds of phorate and disulfoton, two highly toxic organophosphate insecticides had entered agricultural insect control with a vengeance, but malathion use had dropped more than 1.5 million pounds to 3.6 million pounds, and diazinon use had also fallen (by more than 2 million pounds to 3.2 million pounds). Disulfoton stood out as a widely used systemic insecticide; when it was applied to soil, plant roots took up the insecticide and transferred it to all parts of the plant. It was particularly effective against sucking insects, such as aphids, leafhoppers, and thrips, while leaving predators and pollinators unharmed, for the most part. Having said that, disulfoton, which bound to soil, was highly toxic to aquatic organisms, fish, birds, and other wildlife.

PESTICIDES AFTER THE DDT BAN

The ban on DDT and other organochlorines took effect in 1973. After paraffinic oil (60 million pounds) and the organochlorine, toxaphene (34 million pounds), the other top ten insecticides used by farmers were organophosphates and carbamates for total applications of more than 85 million pounds. Toxaphene topped the list of synthetic insecticides in 1976. Farmers typically applied this chlorinated hydrocarbon in mixtures with methyl parathion for insect control in cotton.[34] Although methyl parathion and carbaryl/Sevin use slid to 23 million pounds and less than 16 million pounds, respectively, ethyl parathion use remained fairly constant at about 9 million pounds. Meanwhile, the application of phorate and disulfoton both rose to nearly 7 million pounds each. Like aldrin, both of these chemicals were applied to corn, and the increase more than offset the decline (by 7 million pounds) in the use of aldrin. Two new (and highly toxic) organophosphate insecticides, EPN (6 million pounds) and fonofos (5 million pounds), appeared in the list of the top ten insecticides used in agriculture. Economic entomologists compared EPN to parathion, but it was more persistent. Farmers applied fonofos, also highly toxic, to corn. Shifting the focus to acres rather than pounds, the decline in acres treated with aldrin and bufencarb (a total of more than 11 million acres) was more than offset by acres treated with carbofuran, fonofos, phorate, and terbofos (more than 16 million).[35]

By 1982, use of organochlorines had dropped to such an extent that only toxaphene remained among the top ten insecticides by weight (6.6 million pounds). Several factors explain the precipitous fall in toxaphene applications (from more than 34 million pounds in 1976 to less than 7 million in 1982). First, the organochlorine (usually paired with methyl parathion) was becoming less effective due to the development of resistance in insects. Second, toxaphene use dropped as a result of the replacement of toxaphene and methyl parathion, typically applied to cotton in 1976 at the rate of 3 to 9 pounds per acre, with synthetic pyrethroids (fenvalerate and permethrin), which were applied at the rate of one half pound per acre in 1982. Finally, and perhaps most significant, the EPA banned the use of toxaphene in 1982.[36] With the exception of paraffinic oil (50 million pounds), organophosphate insecticides dominated the list of insecticides applied at rate of more than 500,000 pounds per year. With that in mind, agricultural use of most of the well-known insecticides had fallen since 1976. Use of methyl parathion dropped more than

PESTICIDES AFTER THE DDT BAN

50 percent. Carbaryl and ethyl parathion applications fell more than 6 million pounds and nearly 3 million pounds, respectively. However, as in previous years, reductions were offset by significant increases in the use of such insecticides as terbufos (up 6 million pounds to 8.6 million pounds) and chlorpyrifos (5.4 million pounds). Registered in 1965, chlorpyrifos was applied to a variety of food and feed crops in agriculture especially corn. It was also popular for insect control on golf courses, wood treatment (nonstructural), and against adult mosquitoes. Over time, it became clear that single applications of chlorpyrifos posed risks to small mammals, birds, fish, and aquatic invertebrates for most outdoor uses. Risks to wildlife increased with multiple applications.[37] Like many other organophosphate insecticides, terbufos was extremely toxic to birds, fish, and aquatic invertebrates. Farmers used it as an insecticide and nematicide on corn, sugar beets, and grain sorghum. Carbofuran and fonofos applications remained fairly consistent between 1976 and 1982. By 1982, some cotton farmers had begun to apply two synthetic pyrethroids in significant quantities: permethrin (1.5 million pounds) and fenvalerate (just under 1.3 million pounds). These ongoing trends are illustrated by the data in table 3.

The year 1989 marked a major milestone. Aside from paraffinic oil (for which applications had fallen to 35 million pounds), organophosphate insecticides accounted for the rest of the insecticides in the top twenty agricultural insecticides applied for agriculture (for total applications that exceeded 90 million pounds). Organophosphates represented all but three of the insecticides of which more than 500,000 pounds were applied in 1989 (more than 100 million pounds in total). Specifically, agricultural use of chloripyrifos and terbufos rose to 11.3 million pounds and 10.2 million pounds, respectively. Malathion use rose to 6.3 million pounds, mainly because it was used in the program to eradicate cotton boll weevil.[38] Meanwhile, applications of methyl parathion, carbofuran, and fonofos dropped considerably. Ethyl parathion and phorate use remained fairly consistent with 1982.

Although 90 million pounds of potent insecticides is a vast quantity of chemicals, the number is misleading on several levels. First, it includes only those pesticides used in U.S. agriculture at a rate of 500,000 pounds per year or more in 1989. There were many other organophosphates used widely for which estimates fell short of 500,000 pounds.

203

Table 3
Top Insecticides, 1966–1982 (1,000 pounds active ingredient/year)

1966			1971		
Toxaphene	OC	34,605	Oil		73,950
DDT	OC	27,004	Toxaphene	OC	37,464
Aldrin	OC	14,761	Methyl parathion	OP	27,563
Carbaryl	CB	12,392	Carbaryl	CB	17,838
Paraffinic oil		11,419	DDT	OC	14,324
Ethyl parathion	OP	8,452	Ethyl parathion	OP	9,481
Methyl parathion	OP	8,002	Aldrin	OC	7,928
Diazinon	OP	5,605	Phorate	OP	4,178
Malathion	OP	5,218	Disulfoton	OP	4,079
TDE	OC	2,896	Bufencarb	CB	3,606
Methoxychlor	OC	2,578	Malathion	OP	3,602
Strobane	OC	2,016	Diazinon	OP	3,167
Ethion	OP	2,007	Methoxychlor	OC	3,012
Disulfoton	OP	1,952	Carbofuran	CB	2,854
Dicrotophos	OP	1,857	Azinphos-methyl	OP	2,654
Heptachlor	OC	1,536	Ethion	OP	2,326
Azinphos-methyl	OP	1,474	Chlordane	OC	1,890
Trichlorfon	OP	1,060	Endrin	OC	1,427
Endosulfan	OC	791	Heptachlor	OC	1,211
Endrin	OC	751	Methomyl	CB	1,077
Dieldrin	OC	724	Endosulfan	OC	882
Lindane	OC	704	Dicrotophos	OP	807
Chlordane	OC	526	Lindane	OC	650
Mexacarbate	CB	502	Trichlorfon	OP	617

Table 3 (*continued*)

1976			1982		
Oil		60,000	Oil		50,000
Toxaphene	OC	34,178	Carbofuran	CB	12,300
Methyl parathion	OP	23,350	Methyl parathion	OP	11,335
Carbaryl	CB	15,829	Carbaryl	CB	9,984
Carbofuran	CB	11,623	Terbufos	OP	8,632
Ethyl parathion	OP	9,268	Toxaphene	OC	6,596
Phorate	OP	6,957	Ethyl parathion	OP	6,384
Disulfoton	OP	6,873	Fonofos	OP	5,486
EPN	OP	6,249	Chlorpyrifos	OP	5,412
Fonofos	OP	5,008	Phorate	OP	5,379
Methoxychlor	OC	4,057	Methomyl	CB	4,253
Malathion	OP	3,936	Ethoprop	OP	2,907
Methomyl	CB	3,417	Malathion	OP	2,521
Azinphos methyl	OP	2,644	Disulfoton	OP	2,443
Ethion	OP	2,639	Azinphos methyl	OP	2,274
Terbufos	OP	2,492	Aldicarb	CB	2,271
Diazinon	OP	2,470	Diazinon	OP	2,114
Chlordane	OC	2,116	Permethrin	SP	1,475
Monocrotophos	OP	1,917	Dimethoate	OP	1,419
Heptachlor	OC	1,667	EPN	OP	1,373
Endosulfan	OC	1,653	Mevinphos	OP	1,277
Ethoprop	OP	1,148	Fenvalerate	SP	1,273
Aldrin	OC	945	Ethion	OP	1,250
Trichlorfon	OP	932	Acephate	OP	1,173
Endrin	OC	866	Endosulfan	OC	977
Fensulfothion	OP	748	Methamidophos	OP	942
Acephate	OP	588	Phosmet	OP	903
Dimethoate	OP	583	Sulprofos	OP	794
Phosmet	OP	523	Monocrotophos	OP	761
			Naled	OP	745
			Oxamyl	CB	667

Source: Figures are based on data in Leonard P. Gianessi, "U.S. Pesticide Use Trends: 1966–1989" (Washington, D.C.: Resources for the Future: 1992).

Note: OC = organochlorine or chlorinated hydrocarbon; OP = organophosphate; CB = carbamate; SP = synthetic pyrethroid; oil = paraffinic oil.

Second, the calculation includes only the active ingredients, for example, methyl parathion. Most insecticides require some sort of delivery medium, which are not included in calculations of pesticide use. Often the substances utilized for distributing the active ingredients add to the toxicological profile. Yet another source of underestimation stems from the simple fact that only a portion of pesticides are directed toward agriculture. Just as chemical companies sold chlorinated hydrocarbons widely, they also marketed organophosphates to consumers, who depended on them to replace banned chlorinated hydrocarbons. In addition, municipal public health organizations sprayed extensively for insect control. Just as farmers sought replacements for organochlorines like DDT, public health officials looked to organophosphate insecticides to fill the gap left by the ban on so many chlorinated hydrocarbons. Suffice it to say, 90 million pounds was a very conservative estimate for the quantity of insecticides deployed in 1989.

The details of the quantities and applications of insecticides in the decades after *Silent Spring* reveal several trends. First, as DDT and other chlorinated hydrocarbons underwent legislative and regulatory scrutiny, farmers turned to alternatives; first to another chlorinated hydrocarbon, toxaphene, but increasingly to organophosphates. Second, by 1976, organophosphates dominated insecticides in agricultural use. Finally, despite bans on DDT and other chlorinated hydrocarbons, total insecticide use remained at high levels, which suggests that most farmers continued to view insecticides as a major part of their effort to produce crops.

After twenty-five years of considerable variability in the type and quantity of particular insecticides from 1966 to 1989, a few specific pesticides emerged as the preferred tools for insect control in agriculture. In 1992, 1997, and 2002, chlorpyrifos, terbufos, and methyl parathion topped the list of insecticides most used in agriculture. Several carbamates (carbaryl/Sevin, carbofuran, and aldicarb) remained popular. Demand remained high despite the initiation of a comprehensive review of the organophosphate and carbamate insecticides by EPA (see table 4).

Meanwhile, the link between pesticides and cancer seemed evermore tenuous. In 1988, the Council on Scientific Affairs (CSA) of the American Medical Association reviewed the link between pesticides and cancer. After acknowledging the challenges of establishing such a link and the problems with animal models, the authors concluded that acute tox-

Table 4

Top Insecticides, 1989–2002 (1,000 pounds active ingredient/year)

1989			1992		
Oil		35,000	Oil		51,102
Chlorpyrifos	OP	11,300	Chlorpyrifos	OP	14,765
Terbufos	OP	10,246	Terbufos	OP	8,690
Carbaryl	CB	8,616	Methyl parathion	OP	5,962
Methyl parathion	OP	7,652	Carbofuran	CB	5,101
Carbofuran	CB	7,156	Carbaryl	CB	4,543
Malathion	OP	6,327	Phorate	OP	4,453
Ethyl parathion	OP	6,030	Cryolite	FL	4,053
Phorate	OP	5,329	Aldicarb	CB	4,022
Aldicarb	CB	5,317	Propargite	OS	3,628
Dimethoate	OP	4,250	Acephate	OP	3,390
Fonofos	OP	3,220	Malathion	OP	3,378
Azinphos-methyl	OP	3,000	Fonofos	OP	3,234
Acephate	OP	2,500	Methomyl	OP	2,755
Ethoprop	OP	2,500	Dimethoate	OP	2,619
Methomyl	CB	2,345	Azinphos-methyl	OP	2,549
Disulfoton	OP	2,023	Ethyl parathion	OP	2,318
Diazinon	OP	1,847	Profenofos	OP	2,063
Ethion	OP	1,350	Disulfoton	OP	1,807
Methamidophos	OP	1,259	Endosulfan	OC	1,797
Endosulfan	OC	1,100	Thiodicarb	CB	1,706
Fenvalerate	SP	1,000	Ethoprop	OP	1,450
Phosmet	OP	1,000	Dicofol	OC	1,392
EPN	OP	975	Diazinon	OP	1,266
Thiodicarb	CB	950	Methamidophos	OP	1,088
Profenofos	OP	777	Permethrin	SP	1,069
Mevinphos	OP	757	Ethion	OP	991
Phosalone	OP	750	Oxamyl	OP	946
Methidathion	OP	600	Phosmet	OP	941
Naled	OP	600	Sulprofos	OP	852
Trichlorfon	OP	568	Dicrotophos	OP	666
Dicrotophos	OP	550	Fenamiphos	OP	615
Monocrotophos	OP	550			
Cypermethrin	SP	500			
Esfenvalerate	SP	500			
Permethrin	SP	450			

(*continued*)

Table 4   (*continued*)

1997			2002		
Oil		102,337	Oil		91,606
Chlorpyrifos	OP	13,464	Chlorpyrifos	OP	8,481
Terbufos	OP	6,516	Malathion	OP	5,132
Methyl Parathion	OP	5,917	Aldicarb	CB	3,419
Malathion	OP	5,810	Terbufos	OP	3,363
Carbaryl	CB	4,858	Carbaryl	CB	2,986
Aldicarb	CB	4,278	Acephate	OP	2,525
Carbofuran	CB	3,398	Methyl parathion	OP	2,148
Phorate	OP	3,218	Kaolin		1,690
Cryolite	FL	2,560	Phosmet	OP	1,495
Propargite	OS	2,539	Propargite	OS	1,407
Acephate	OP	2,462	Dimethoate	OP	1,346
Azinphos-methyl	OP	2,091	Azinphos-methyl	OP	1,224
Methomyl	OP	1,997	Phorate	OP	1,197
Dimethoate	OP	1,897	Cryolite	FL	1,102
Endosulfan	OC	1,601	Carbofuran	CB	1,015
Phosmet	OP	1,333	Dicrotophos	OP	980
Disulfoton	OP	1,196	Methomyl	OP	918
Permethrin	SP	1,066	Endosulfan	OC	868
Ethoprop	OP	1,011	Diazinon	OP	858
Methamidophos	OP	966	Oxamyl	OP	748
Oxamyl	OP	939	Tefluthrin	SP	630
Diazinon	OP	918	Permethrin	SP	586
Profenofos	OP	880	Tebupirimphos	OP	538
Thiodicarb	CB	821			
Dicofol	OC	787			
Fenamiphos	OP	727			
Naled	OP	605			
Tefluthrin	SP	577			
Ethyl parathion	OP	529			
Ethion	OP	505			

Source: The 1989 figures are based on data in Leonard P. Gianessi, "U.S. Pesticide Use Trends: 1966–1989" (Washington, D.C.: Resources for the Future, 1992); 1992–2002 figures are based on data in Leonard P. Gianessi and Nathan Reigner, "Pesticide Use U.S. Crop Production: 2002" (Washington, D.C.: Croplife Foundation, 2006).

Note: OC = organochlorine or chlorinated hydrocarbon; OP = organophosphate; CB = carbamate; SP = synthetic pyrethroid; oil = paraffinic oil; FL = fluoridated insecticide.

icity was the primary hazard of pesticide exposures and that no pesticides had been proven to be carcinogenic, despite evidence of carcinogenicity in animals: "A large number of pesticidal compounds have shown evidence of genotoxicity or carcinogenicity in animals and in vitro screening tests, but no pesticides—except arsenic and vinyl chloride (once used as an aerosol propellant)—have been proved definitely to be carcinogenic in man. Epidemiological studies offer only conjectural evidence at best that pesticides may be carcinogenic."[39] Nevertheless, the CSA recommended that the AMA urge the EPA to intensify its efforts at pesticide reregistration to determine the long-term health effects of pesticides especially carcinogenicity.

For the most part, organophosphate insecticides were not associated with carcinogenicity, so they passed through the screen that was the regulatory emphasis on cancer. Since they typically did not bioaccumulate in the environment, they avoided one of the chief drawbacks of the organochlorines. Lost in these toxicological analyses was the damage that organophosphate insecticides wrought to humans and wildlife directly in the form of acute toxicity. As we have seen, with the exception of malathion, organophosphates were moderately to highly toxic to humans and wildlife, especially birds, fish, aquatic organisms, and nontarget insects, including bees. To a degree that would have shocked and disappointed Carson, the "road traveled" was flooded with highly toxic organophosphate insecticides, which she had identified as some of the most toxic chemicals known to man.

In the mid-1990s Theo Colborn, a World Wildlife Fund research scientist, pieced together evidence that pointed to a strikingly different concern. Drawing on hundreds of published studies, Colborn argued that environmental chemicals caused endocrine disruption in a wide range of animals and humans.[40] Such a finding squared well with one of Rachel Carson's greatest concerns in *Silent Spring,* namely, the decline of topline predators as a result of eggshell thinning due to the bioaccumulation of chlorinated hydrocarbons like DDT (and for that matter PCBs). Endocrine disruption, as Langston has shown, was a neglected element of risk assessment, with serious consequences.[41]

Neither cancer nor endocrine disruption was on Kenneth DuBois's mind when he expressed concern about the imminent ban on DDT. DuBois's concern was neurotoxicity. With the DDT ban, DuBois worried

209

PESTICIDES AFTER THE DDT BAN

that farmers and public health officials would turn to organophosphates to control insects, thereby exposing farm workers and others to extremely toxic chemicals. DDT was banned as were other chlorinated hydrocarbons, but it was many years before any of the organophosphates underwent the kind of scrutiny that could support a move to phase them out. In fact, most organophosphates were still in use as of 1996. As Du-Bois and others feared, organophosphate insecticides replaced DDT for many general uses. Between 1964 and 1994, pesticide use in the United States doubled from 500 million pounds to over 1 billion pounds.[42] As we saw above, more than half of the pesticides in use through the 1990s were organophosphates.

Wildlife continued to perish at phenomenal rates largely due to exposures to organophosphates. In 1997, *Audubon* (magazine) reported that more than 67 million birds were dying annually as a result of pesticide poisoning in the United States.[43] Monocrotophos is one example of an organophosphate particularly toxic to birds. It was initially registered in the United States in 1965. Certain scientists believe that monocrotophos has been responsible for more avian mortality incidents than any other pesticide since 1965. EPA canceled all registered use of this chemical in 1991, and the largest U.S. manufacturer voluntarily began to phase out its production, but monocrotophos and many other organophosphates are still in use internationally, where they pose significant risks to humans and wildlife. One extreme example arose in Argentina in 1996 when thousands of Swainson's hawks died at their core wintering site after monocrotophos spraying. As many as three thousand hawks at one site and perhaps twenty thousand in all perished.[44] Argentina subsequently banned monocrotophos from agricultural use.

As DuBois predicted, urban and suburban use of pesticides put humans and wildlife seriously at risk. Until its ban took effect in 2001, Americans used 6 million pounds annually of diazinon, 70 percent of it used by homeowners and professional applicators for structural and lawn pest control around residences and public buildings. Diazinon applications have caused the second largest number of total bird deaths of any pesticide.[45] Birds are not alone in their susceptibility to organophosphates, although one of the legacies of *Silent Spring* is a particular public sensitivity to avian mortality. Populations of mammals, fish, reptiles,

PESTICIDES AFTER THE DDT BAN

amphibians, as well as beneficial and nontarget insects, suffer from exposure to various pesticides, herbicides, and fungicides.

As environmental historian Linda Nash elegantly argued, farm workers regularly faced exposures to these substances, in violation of state and federal regulations and at levels that can inhibit cholinesterase. In *Our Children's Toxic Legacy,* Wargo noted that the other group most at risk is children who consume more of the liquids, fruits, and vegetables that may carry organophosphates. Children may also encounter organophosphates applied indoors. Animal studies continue to sharpen scientists' understanding of the risks posed by organophosphates.[46] For example, there is substantial toxicological evidence that repeated low-level exposure to organophosphate pesticides may affect neurodevelopment and growth in developing animals.[47] At the peak of use during the 1990s, there may have been as many as ten thousand cases of organophosphate poisoning annually in the United States alone. Internationally, organophosphates still pose grave risks to children and farmworkers. In July 2013, twenty-three Indian children (aged five to twelve) died and dozens more were sickened after consuming free school lunches accidentally contaminated with the organophosphate monocrotophos at a school in the state of Bihar.[48] Similar cases have been reported in China and Ecuador.

In 1996, President Bill Clinton signed the Food Quality Protection Act (FQPA), which amended both FIFRA (1947) and the FFDCA (1938). The FQPA required the EPA to reassess all food tolerances established before August 3, 1996, giving priority to those pesticides posing the greatest risk. This act compelled the EPA to conduct an extensive cumulative risk assessment of the organophosphates. The forty or so organophosphates in use were among the first chemicals the EPA reviewed, followed by the other group of chemicals that induce cholinesterase inhibition: the carbamates. The deadline for the EPA to complete its review of all tolerances was August 2006.[49] Although progress was slow, the EPA announced the phaseout of chlorpyrifos and diazinon based on toxicity and the risk they posed to children through contaminated food and drinking water, as well as their threat to birds and other wildlife.[50] At the close of the 2006 cumulative risk assessment, the EPA announced the cancellation of many other organophosphates and carbamates.

211

Recent studies have challenged conventional wisdom regarding the toxicity of organophosphates. Since DuBois and other researchers determined the very high neurotoxicity of organophosphates at very small exposure levels, few researchers pursued studies of chronic exposures to organophosphates. In 2011, three different research teams published on the neurotoxic effects of organophosphates in *Environmental Health Perspectives,* one of the leading journals of environmental health. Drawing on data from prenatal exposures to organophosphates resulting from spraying crops in Salinas, California, and from spraying roaches in New York City apartments, the studies reached similar conclusions: prenatal exposures to organophosphate insecticides impaired intellectual development in children. The researchers at the Mailman School of Public Health, Columbia University, found IQ declines of 1.4 percent on average in children with prenatal exposures to chlorpyrifos.[51] Another group studying New York residents discovered that exposure to organophosphate pesticides was negatively associated with cognitive development, particularly perceptual reasoning.[52] Even more striking, a third study of children exposed to organophosphates in agricultural areas in Salinas, California, showed that those in the group that suffered the highest exposures exhibited an average deficit of seven IQ points compared with children in the lowest exposure group.[53] Even more surprising, yet another study revealed a possible association between low level exposures to organophosphate flame retardants (another common use of the chemicals) and two effects: altered hormone levels and decreased semen quality in men.[54] Suffice it to say, current research has broadened the toxicological profile of organophosphates.

The story of organophosphates and environmental risk continued to unfold up to and beyond 2006. Neither science nor regulation came to terms with this group of highly toxic chemicals until decades after restrictions and bans were placed on DDT and other chlorinated hydrocarbons. At the very least, Rachel Carson's warning inspired grassroots environmental activism against DDT and other chlorinated hydrocarbons. The ban on DDT and protection under the Endangered Species Act (1973) has contributed to the recovery of numerous species of wildlife, most notably bald eagles, peregrines, brown pelicans, and ospreys. But more than three decades passed before the EPA completed its cumulative review of organophosphates. For most of that period, risk assessment focused on cancer, which reflected a public health priority dating back to World War II. Even when risk assessment broadened to include endocrine disruptors in the 1990s, most organophosphates slipped through the regulatory net. Only with the passage of the Food Quality Protection Act in 1996 and the cumulative review of organophosphates in 2006 did the organophosphates, some of the most toxic chemicals available, receive the careful scrutiny that led to the removal of these pesticides from the market. Introduced in the same time frame as DDT, the organophosphate insecticides, which were highly toxic nerve toxins, slipped through the cracks in the regulatory frameworks established in the wake of *Silent Spring*. Several recent studies have significantly broadened the toxicological profile of organophosphates to include cognitive impairment in children exposed prenatally and possible endocrine effects.

CHAPTER 8

Roads Taken

What lessons might we draw from a century of pesticides, a hundred years of risk assessment? Despite the prodigious efforts of scientists, regulators, and legislators, simple solutions have not emerged. As we have seen, as the scale of agriculture developed to industrial proportions, farmers increasingly relied on chemical inputs—and particularly insecticides—to control crop damage as a result of insect infestations. A similar process of upscaling placed incredible distance between producers (of meat and other products) and consumers. When American consumers and legislators discovered the health risks that such distances entailed, Congress passed the 1906 Pure Food and Drug Act. The Insecticide Act of 1910 followed a few years later. Like the PFDA, the Insecticide Act was essentially a law for truth in labeling. Although these laws were groundbreaking at the time of passage, cracks soon appeared in the legislation that left consumers exposed to toxic substances. Several egregious cases that left thousands injured and worse and reports in popular books and articles accelerated the progress toward revised legislation in the form of the 1938 Federal Food, Drug, and Cosmetic Act. The efforts of University of Chicago pharmacologist E. M. K. Geiling and FDA scientists to develop rigorous quantitative methods revealed the value of toxicology in characterizing risks of chemicals in the marketplace. Contemporaneously, Alice Hamilton, researchers with the Harvard Lead Study, and W. C. Hueper, among others, initiated the study of industrial disease with important implications for the study of toxicology and carcinogenicity.

214

The development and deployment of DDT during World War II is a story of remarkable technological advancement. Employing techniques of toxicology, scientists conducted numerous tests of the new insecticide. Alongside the many reports of DDT's spectacular success against numerous target insects, toxicology evaluations indicated relatively low toxicity in lab animals, field studies, and even humans. Enthusiastic accounts swamped concerns regarding threats to nontarget insects, the development of resistance, and uncertain chronic toxicity.

The Toxicity Laboratory at the University of Chicago emerged as a major center for research and development in this new field study under Geiling's deft leadership. The various research programs at the Tox Lab produced refined methods of toxicology, including the calculation of joint toxicity from the study of antimalarial drug therapies, the determination of toxicity of minute doses of drugs utilizing radioisotopes, and the calculation of LD_{50}s for thousands of chemicals. Kenneth DuBois, in tandem with FDA researchers, developed the toxicological profile for some of the organophosphate insecticides, which caused cholinesterase inhibition in organisms, including humans, and could potentiate in combination with other pesticides from the class. With the possible exception of malathion, organophosphates as a class were among the most toxic of chemicals.

As scientists at the Tox Lab and the FDA grappled with the new insecticides and developed new techniques, Congress began to consider the implications of the new chemicals for existing legislation. Congressional hearings explored the risks and benefits of pesticides, but the hearings were shrouded in the mists of scientific uncertainty. Repeatedly, committee members demanded a clear statement of the risks of DDT and were frustrated and angered as witness after witness failed to provide one. Nevertheless, Congress passed significant legislation in the form of FIFRA (1947), the Miller Amendment, and the Delaney Clause, which banned the use of cancer-causing agents in the production of foodstuffs. The aminotriazole cranberry scare in 1959 caused the FDA to apply the Delaney Clause and seize contaminated crops. Meanwhile, through the efforts of Tox Lab and FDA scientists, toxicology had begun to coalesce as an independent field of study and as a profession that, Geiling argued, should acknowledge its responsibility to the public.

Despite Geiling's ambitions for toxicologists, it was a science writer who publicized the problems pesticides presented for humans, wildlife, and ecosystems. In *Silent Spring*, published in 1962, Rachel Carson eloquently synthesized the research of toxicologists, ecologists, and doctors to portray the risks of indiscriminate use of insecticides. Many of her examples involved DDT and other chlorinated hydrocarbons, but she also noted the considerable risks associated with the use of organophosphates. Carson's message reached millions of Americans when CBS aired a program dedicated to *Silent Spring*. Her book inspired further hearings when President Kennedy directed the PSAC to investigate effects of insecticides. Congressional hearings followed. Again, witnesses testified on the benefits and risks of pesticides. Carson herself distilled the message of *Silent Spring* into a call for specific research. As in other hearings, the committee listened to testimony from many experts regarding the toxicity of organophosphates, extrapolating what they could from the no-effect level and the multiyear process of bringing a new insecticide to market. However, these hearings did little to change pesticide legislation.

Late in 1972, after extended litigation on several fronts, the EPA banned DDT and other chlorinated hydrocarbons. Banning the related compounds aldrin and dieldrin provided the first test cases of the recently passed FEPCA. Received wisdom marked the DDT ban as one of the great achievements of the environmental movement in the United States, and during the ensuing decades the gradual but pronounced recovery of populations of bald eagles, ospreys, peregrines, brown pelicans, and other wildlife confirmed the sense of accomplishment. The effects of DDT were pernicious, and banning the chemical in the U.S. mitigated risks to wildlife. But where would farmers turn when they could no longer spray DDT to control the insects that threatened their crops? DuBois, the Chicago toxicologist, worried that highly toxic organophosphates would become the insecticides of choice. In 1973, an English research lab synthesized pyrethroids, which were similar to the natural insecticide pyrethrum in chemical structure and insecticidal action. Synthetic pyrethroids combined high toxicity to insects with extremely low toxicity to mammals. But between 1966 and 1989, as DuBois had feared, organophosphates emerged as the most prolific

ROADS TAKEN

insecticides deployed in American agriculture. Despite their high toxicity, organophosphates did not bioaccumulate in the environment, nor were they considered to be carcinogenic. Nevertheless, before 2006, when most uses of these insecticides were banned by the EPA, they caused thousands of poisonings among farm workers and huge mortality in wildlife inadvertently exposed. In the meantime, the University of Chicago Tox Lab scientists had dispersed to research universities, medical centers, and federal regulatory agencies. The roots laid down at the Tox Lab continue to feed the science of toxicology.

The lessons from the story of pesticides and toxicology are many. First, as we have seen repeatedly, risk and benefit form a tight helix in the case of insecticides. The knowledge that insects carry disease and destroy crops neglects the considerable nuisance they create as we try to remake landscapes to exclude them. Human efforts to control insects almost certainly predate history, and natural insecticides like pyrethrum offered the promise of control, but with the introduction of heavy metal insecticides in the late nineteenth century, farmers, already in the process of expanding agriculture to an industrial scale, discovered what seemed to be a magic bullet. The many millions of pounds of lead arsenate deployed in agriculture before World War II provided a clear indication that farmers had wholly embraced synthetic insecticides. During and after the war, DDT offered an apparently safer option with no obvious drawbacks, especially when compared with the far more toxic organophosphates, which were developed and introduced in the same time frame. Early testing, though extensive, failed to detect the most pernicious of the effects associated with DDT: namely, concentration in organisms and the environment resulting in endocrine disruption, particularly in topline predators, such as bald eagles, peregrines, ospreys, and brown pelicans. The risks of organophosphates, however, were absolutely clear to toxicologists shortly after the novel toxins arrived at the Tox Lab and the FDA.

The second lesson is that there is a danger of focusing on any particular element of toxicity in weighing the risks and benefits of insecticides. DDT handily replaced the arsenates because it destroyed insects without the obvious toxicity of lead and arsenic and without damaging crops. In *Silent Spring,* Carson revealed the considerable dangers of chlorinated

ROADS TAKEN

hydrocarbons and organophosphates. DDT and chlorinated hydrocarbons accumulated in ecosystems and organisms; the highly toxic organophosphates killed outright. If pressed, as she was during congressional hearings, Carson would have recommended further research into the effects of all pesticides, and she would have strongly urged reducing dependence on all chemical insecticides, certainly chlorinated hydrocarbons *and* organophosphates. The fact that organophosphates did not bioaccumulate would not offset the extraordinary risks they posed to humans and wildlife, a point that DuBois stressed. Recent studies revealing cognitive effects associated with prenatal exposures to organophosphates as well as possible endocrine effects serve to underscore this point.

Third, pesticides do not offer simple solutions. Replacing arsenates with DDT did not completely solve the problems created by arsenates. The unintended consequences of banning DDT continue to reverberate in agriculture and public health in America and throughout the world. The toxicologists and other scientists in this story have indicated that the toxicology of insecticides and other chemicals is complex. Solutions require a comparable sophistication. The several court cases that challenged USDA's DDT spraying campaigns signaled the need for further regulation of DDT. Comprehensive regulation would have taken organophosphates into account as well. Instead, most organophosphates remained in use for another thirty years with hundreds of millions of pounds in annual agricultural applications. Surely, the dominance of organophosphates in agriculture represents one of the most tragic ironies of the DDT ban.

But we still have not answered the obvious question: namely, why did highly toxic organophosphates replace chlorinated hydrocarbons in American agriculture? In *Silent Spring*, Carson seemed prescient in that she addressed risks of both chlorinated hydrocarbons and organophosphates, but, as I have shown, in developing her case Carson drew on the research and testimony of the toxicologists. Over the course of many hearings, Congress heard testimony regarding the toxicity of the organophosphates.

DDT was similar to the arsenates that preceded it for its persistence in the environment. Regulation proceeded along similar lines. By the early 1970s, there were thousands of products containing DDT, which had not been a proprietary pesticide since its release after World War II.

218

ROADS TAKEN

Banning DDT did not place a burden on any one chemical company in the way that banning a proprietary pesticide would have. Most, if not all, of the organophosphates were proprietary, which is to say individual chemicals were associated with specific chemical companies. Put more simply, in the aftermath of the DDT ban, the chemical companies producing proprietary organophosphates realized significant profits. Ironically, under FEPCA chemical companies demanded indemnification for stocks of pesticides banned under the legislation.

During the DDT era, environmental toxicology relied on two key measures to determine the relative safety of pesticides: carcinogenicity and environmental persistence. In the case of organophosphates, with such low LD_{50}s (high toxicities), carcinogenic and reproductive effects were obscured by symptoms of acute toxicity, such as convulsions, paralysis, and even death. Only recently have scientists tracked reproductive and neurological effects of low dose exposures to organophosphates. Organophosphates decompose rapidly in most environmental systems and thus slipped through the regulatory screens put in place by FIFRA and FEPCA, which focused on imminent hazard to humans and the environment and focused on restricting carcinogens and persistent pollutants. Again, recent studies have revealed the harmful effects of organophosphates in cognitive development and endocrine disruption. This research revealed significant gaps in pesticide legislation.

When Carson referred to certain pesticides as "biocides" she certainly had organophosphates in mind. Recall that when scientists synthesized pyrethroids, they compared their toxicity to parathion, one of the most toxic pesticides (to all organisms). Since their development as nerve gasses during World War II, scientists were well aware of the toxicity of organophosphates to all organisms. Toxicologists at the Tox Lab and the FDA provided a clear picture of toxicity of organophosphates as cholinesterase inhibitors. In *Silent Spring*, Carson animated the dangers that organophosphates posed to wildlife and agriculture laborers. Numerous witnesses testified to the toxicity of organophosphates in hearings held at the state and federal level, particularly during PSAC and subsequent congressional hearings. Few classes of chemicals provided such a clear and consistent picture of extreme risk as the organophosphates (with malathion a notable exception).

219

ROADS TAKEN

Despite Carson's careful account and thoughtful recommendations, legislators, regulators, and the public focused more narrowly on the persistent chlorinated hydrocarbons in the aftermath of *Silent Spring*. Note the parallel with the regulation of the pesticides that preceded DDT, the arsenates, which were also pollutants that bioaccumulated in the environment. The ban on DDT in 1972 left farmers in search of a technological fix for the problem of insect infestations, which led them to the organophosphates, with the guidance of the USDA and the chemical industry. Although scientists developed synthetic pyrethroids shortly after the ban, their commercial development lagged and they remained prohibitively expensive.

My attempt to understand the development of pesticides and toxicology through the sources to *Silent Spring* has revealed the tragic irony of legislation and pesticide use in the aftermath of the book's publication. Along with a generation of toxicologists, Carson and her careful readers knew that the organophosphates posed an equivalent (and potentially greater) risk to humans and wildlife. Yet most of the organophosphates remained on the market and dominated agricultural pesticide applications in the United States until the EPA completed its comprehensive review in 2006. To this day, organophosphates are among the most widely used pesticides in the world, with tragic consequences for farm workers, children, and wildlife populations alike.

220

Epilogue

Risk, Benefit, and Uncertainty

Even before the EPA began its review of the organophosphates, a new class of insecticides had joined the ranks of agricultural insecticides. Since the 1950s, scientists have attempted to synthesize compounds like the naturally occurring insecticide nicotine. Izuru Yamamoto at the Tokyo University of Agriculture coined the term "nicotinoid" for nicotine and related insecticidal compounds. Chemists first synthesized promising nicotinoids during the 1970s, but the initial compounds were unstable in light and thus unviable for development as insecticides. Agricultural chemists working with support from Bayer and Shell successfully developed and patented several "neonicotinoids," also Yamamoto's term, during the 1980s and 1990s. As a class, neonicitinoid insecticides showed promise as systemic insecticides that would be taken up by crops like some of the organophosphates.[1] Toxicological analysis showed that the new insecticides were highly toxic to insects and minimally toxic to mammals. For example the LD_{50} in rats for imidacloprid (IMI) was 450 mg/kg, thiacloprid: 640 mg/kg, and clothianidin: > 5,000 mg/kg. However, both imidacloprid and thiacloprid indicated much lower LD_{50}s for birds: 31 and 49 mg/kg, respectively.[2]

Agricultural usage of neonicitinoids expanded as they became more widely available, but use exploded when the EPA cancelled the registrations of many of the organophosphates. In 2013, neonicotinoids surpassed organophosphates as the most widely used insecticides in the

221

EPILOGUE

world. Scientists at the American Bird Conservancy recently argued that such widespread usage may spell disaster for birds, particularly those species that favor open grasslands.[3] Like chlorinated hydrocarbons, neonicintinoids persist in soil. They can accumulate in the environment over time. Toxicities to birds compare with the toxicity of organophosphates. And birds are not alone in their vulnerability to neonicitinoids: bees have shown considerable susceptibility to the new insecticides.[4] In addition, there are concerns that neonicitinoids will contaminate groundwater.[5]

Even though neonicitinoids represent a new class of pesticide, or at least a recently synthesized form of an existing chemical (nicotine), their widespread proliferation in agriculture across America and throughout the world seems eerily familiar. Once again, a fog of scientific uncertainty surrounds the most widely used agricultural insecticides in the world. Neonicitinoids account for one-quarter of insecticides used worldwide, with an estimated value of $2.5 billion. The EPA has deemed the neonicitinoid insecticides safe. Yet more and more scientists worry that these chemicals are responsible for ecological disruption and the destruction of populations of birds, bees, and aquatic organisms. Such risks have prompted action on the part of the European Commission, which announced that it would restrict the use of three neonicitinoids (clothianidin, IMI, and thiametoxam) for a period of two years commencing December 1, 2013. Although the EPA reached scientific conclusions similar to those of the European Food Safety Authority regarding the potential for acute effects and uncertainty about chronic risk, it has not elected to restrict use of the neonicitinoids. Although the EPA is currently reviewing the neonicitinoids, for the time being the agency has accepted industry claims that the benefits of the new insecticides significantly outweigh the risks.[6]

It would be foolish to overdraw comparisons between the past and present, yet the similarities speak to our discussion of risk, benefit, and uncertainty. When Rachel Carson penned *Silent Spring,* both organochlorines and organophosphates were widely used in agriculture. Yet uncertainty clouded both science and policy. In a stroke of genius, Carson assembled a range of scientific and anecdotal sources into an impassioned call for reflection on the part of legislators and the public as well as for further investigation by toxicologists and environmental scien-

222

EPILOGUE

tists. Establishing the EPA, the ban on DDT, and the passage of FEPCA all served as critical steps in the management of risk. Despite these and other developments, organophosphate use surged in the decades that followed.

Contrary to Carson's clarion call for reduction in the use of all insecticides, the ban on DDT and other organochlorines initiated a risk-risk trade-off in which agribusiness replaced DDT and the persistent organochlorines with highly toxic organophosphates, like parathion, that threaten the welfare of humans and wildlife despite relatively rapid disintegration in the environment. When Congress enacted FQPA, the EPA launched its comprehensive review of the organophosphates and carbamates, and U.S. restrictions on many of them followed. Nevertheless, neonicitinoids provided agribusiness with substitutes, albeit ones that may contaminate ecosystems and threaten nontarget organisms, including bees and birds. Initial assessments suggest that neonicitinoids pose lower risks to humans and other mammals than the organophosphates and carbamates. As regulators review these chemicals and the risks they pose to ecosystems and wildlife, we should look to *Silent Spring* and a century of pesticides and toxicology for models with which to evaluate novel risks.

NOTES

Abbreviations to Notes

Environmental Health Perspectives	*E.H.P.*
Journal of the American Medical Association	*J.A.M.A.*
Journal of Economic Entomology	*J.E.E.*
Journal of Pharmacology and Experimental Therapeutics	*J.P.E.T.*
Journal of Wildlife Management	*J.W.M.*
Public Health Reports	*P.H.R.*

Chapter 1. Toxicology Emerges in Public Health Crises

1. See William Cronon, *Nature's Metropolis: Chicago and the Great West* (New York: W. W. Norton, 1991), James Harvey Young, *Pure Food: Securing the Federal Food and Drugs Act of 1906* (Princeton: Princeton University Press, 1989), and Oscar E. Anderson, Jr., *The Health of a Nation: Harvey W. Wiley and the Fight for Pure Food* (Chicago: University of Chicago Press, 1958).

2. Adelynne Hiller Whitaker, "Federal Pesticide Legislation in the United States to 1947" (Ph.D. diss., Emory University, 1974), 46.

3. For a thorough study of sulfanilamide and the toxic diethylene glycol, see James Harvey Young, "Sulfanilamide and Diethylene Glycol," in *Chemistry and Modern Society: Historical Essays in Honor of Aaron J. Ihde,* ed. John Parascandola and James C. Whorton (Washington, D.C.: American Chemical Society, 1983). For a brief discussion of the case with respect to the Food, Drug, and Cosmetics Act of 1938, see Whorton, *Before Silent Spring: Pesticides and Public Health in Pre-DDT America* (Princeton: Princeton University Press, 1974). See also Charles O. Jackson, *Food and Drug Legislation in the New Deal* (Princeton: Princeton University Press, 1970), 151–174, Philip J. Hilts, *Protecting America's Health: The FDA, Business, and One Hundred Years of Regulation* (New York: Alfred A. Knopf, 2003), 72–94, and Daniel Carpenter, *Reputation and Power:*

225

NOTES TO PAGES 4–9

Organization Image and Pharmaceutical Regulation at the FDA (Princeton: Princeton University Press, 2010), 85–117.

4. Whorton, *Before Silent Spring*, 5.

5. Ibid., 15–16.

6. Quoted in ibid., 21.

7. See Robert J. Spear, *The Great Gypsy Moth War: A History of the First Campaign to Eradicate the Gypsy Moth, 1890–1901* (Amherst: University of Massachusetts Press, 2005).

8. Whitaker, "Federal Pesticide Legislation in the U.S.," 81.

9. Ibid., 104.

10. Ibid., 107.

11. Christopher Bosso, *Pesticides and Politics: The Life Cycle of a Public Issue* (Pittsburgh: University of Pittsburgh Press, 1987), 48.

12. See Christopher Sellers, *Hazards of the Job: From Industrial Disease to Environmental Health Science* (Chapel Hill: University of North Carolina Press, 1997).

13. See Christian Warren, *Brush with Death: A Social History of Lead Poisoning*, Baltimore: Johns Hopkins University Press, 2000, David Rosner and Gerald Markowitz, *Deadly Dust: Silicosis and the Politics of Occupational Disease in Twentieth-Century America* (Princeton: Princeton University Press, 1991), and David A. Hounshell and John Kenly Smith, Jr., *Science and Corporate Strategy: DuPont R&D, 1902–1980* (Cambridge: Cambridge University Press, 1988).

14. Seller, *Hazards of the Job*, 21–31.

15. Hounshell and Smith, *Science and Corporate Strategy*, 555.

16. See Sellers, *Hazards of the Job*.

17. Ibid., 164.

18. Ibid., 166–172.

19. This section draws on Robert Proctor, *Cancer Wars: How Politics Shapes What We Know and Don't Know About Cancer* (New York: Basic Books, 1995), 36–48, Benjamin Ross and Steven Amter, *The Polluters: The Making of Our Chemically Altered Environment* (Oxford: Oxford University Press, 2010), 59–72, Sellers, *Hazards of the Job*, 221–223, Proctor, "Discovering Environmental Cancer: Wilhelm Hueper, Post-World War II Epidemiology, and the Vanishing Clinician's Eye," *American Journal of Public Health* 87 (11) (November 1997): 1824–1835, and Hounshell and Smith, *Science and Corporate Strategy*, 555–563.

20. Hounshell and Smith, *Science and Corporate Strategy*, 563. See also Ross and Amter, *Polluters*, 61.

21. Proctor, *Cancer Wars*, 40. See also Ross and Amter, *Polluters*, 59–72.

22. Proctor, *Cancer Wars*, 41.

23. Paul B. Dunbar, "Memories of Early Days of Federal Food and Drug Law Enforcement," *Food, Drug, Cosmetic Law Journal* 14 (February 1959), 134. Cited in Jackson, *Food and Drug Legislation*, 3.

24. Jackson, *Food and Drug Legislation*, 4–5.

NOTES TO PAGES 10–19

25. Arthur Kallet and F. J. Schlink, *100,000,000 Guinea Pigs: Dangers in Every-day Foods, Drugs, and Cosmetics* (New York: Grosset and Dunlap, 1933), 4.

26. Ruth deForest Lamb, *American Chamber of Horrors: The Truth about Food and Drugs* (New York: J. J. Little and Ives, 1936), 3.

27. Whorton, *Before Silent Spring*, 81–82.

28. Note: throughout this study, measurements and units will appear as in the original source. For the sake of comparison and consistency, I have inserted metric conversions as appropriate.

29. See Ross and Amter, *Polluters*, 46, and Whitaker, "Federal Pesticide Legislation in the U.S.," 342–343.

30. Ibid.

31. C. N. Myers, Binford Throne, Florence Gustafson, and Jerome Kingsbury, "Significance and Danger of Spray Residue," *Industrial and Engineering Chemistry* (June 1933): 624.

32. Ibid., 625.

33. Kallet and Schlink, *100,000,000 Guinea Pigs*, 57.

34. Whorton, *Before Silent Spring*, 200.

35. Remarkably, enforceable tolerances for lead and arsenic were not set until 1950, by which time DDT and other synthetic insecticides had replaced heavy metal insecticides. See Ross and Amter, *Polluters*, 49–51.

36. This section draws on Daniel E. Rusyniak, R. Brent Furbee, and Robert Pascuzzi, "Historical Neurotoxins: What We Have Learned from Toxins of the Past about Diseases of the Present," *Neurologic Clinics* 23 (2005): 337–352; John P. Morgan and Thomas C. Tulloss, "A Toxicologic Tragedy Mirrored in American Popular Music," *Annals of Internal Medicine* 85 (1976): 804–808; John P. Morgan, "The Jamaica Ginger Paralysis," *J.A.M.A.* 248 (October 15, 1982): 1864–1867; and John Parascandola, "The Public Health Service and Jamaica Ginger Paralysis in the 1930s," *P.H.R.* 110 (May–June 1995): 361–363.

37. Morgan, "Jamaica Ginger Paralysis," 1866.

38. Rusyniak et al., "Historical Neurotoxins," 339.

39. Morgan and Tulloss, "Toxicologic Tragedy," 804–808.

40. Maurice I. Smith, "The Pharmacological Action of Certain Phenol Esters, with Special Reference to the Etiology of So-Called Ginger Paralysis (Second Report)," *P.H.R.* 45 (42) (October 17, 1930): 2518.

41. B. T. Burley, "The 1930 Type of Polyneuritis," *New England Journal of Medicine* 262 (1930): 1139–1142.

42. Cited in Kallet and Schlink, *100,000,000 Guinea Pigs*, 155. Emphasis added.

43. John Pfeiffer, "Sulfanilamide: The Story of a Medical Discovery," *Harper's Magazine* 178 (1939): 387.

44. Ibid.

45. Anon., "Young Roosevelt Saved by New Drug: Doctor Uses Prontylin in Fight on Streptococcus Infection," *New York Times* (December 17, 1936), 1.

NOTES TO PAGES 19–27

46. Waldemar Kaempffert, "The Week in Science: Cause of the Tulsa Deaths," *New York Times* (October 24, 1937), 6.

47. Henry A. Wallace, *Elixir Sulfanilamide: Letter from the Secretary of Agriculture,* in 75th Cong., 2nd Sess., Senate Document 124, Serial 10247 (Washington, D.C.: U.S. Government Printing Office, 1937), 3.

48. Young, "Sulfanilamide and Diethylene Glycol," 108. Though seemingly naïve, Watkins's statement merits further analysis. See below.

49. Morris Fishbein, "Sulfanilamide—A Warning," *J.A.M.A.* 109 (1937): 1128.

50. Ibid.

51. Ibid.

52. Homer A. Ruprecht and I. A. Nelson, "Clinical and Pathologic Observations," *J.A.M.A.* 109 (1937): 1537.

53. Wallace, *Elixir Sulfanilamide,* 4–5, reprinted as Wallace, "Deaths due to Elixir of Sulfanilamide-Massengill. Report of Secretary of Agriculture Submitted in Response to House Resolution 352 of Nov. 18, 1937, and Senate Resolution 194 of Nov. 16, 1937," *J.A.M.A.* 109 (1937): 1986.

54. Ibid.

55. Ibid.

56. Paul Nicholas Leech, "Elixir of Sulfanilamide-Massengill: II," *J.A.M.A.* 109 (1937): 1531.

57. Wallace, *Elixir Sulfanilamide,* 1.

58. E. W. Schoefel, H. R. Kreider, and J. B. Peterson, "Chemical Examination of Elixir of Sulfanilamide-Massengill," *J.A.M.A.* 109 (1937): 1532.

59. E. M. K. Geiling, Julius M. Coon, and E. W. Schoefel, "Preliminary Report of Toxicity Studies on Rats, Rabbits and Dogs Following Ingestion in Divided Doses of Diethylene Glycol, Elixir of Sulfanilamide-Massengill and 'Synthetic' Elixir," *J.A.M.A.* 109 (1937): 1532.

60. Ibid., 1535.

61. Ibid.

62. Ibid.

63. Paul R. Cannon, "Pathologic Effects Following the Ingestion of Diethylene Glycol, Elixir of Sulfanilamide-Massengill, 'Synthetic' Elixir of Sulfanilamide, and Sulfanilamide Alone," *J.A.M.A.* 109 (1937): 1536–1537.

64. See Carpenter, *Reputation and Power,* 76. See also Young, *Pure Food,* and Anderson, *Health of a Nation.*

65. Bosso, *Pesticides and Politics,* 49.

66. Bert J. Vos et al., "History of the U.S. Food and Drug Administration: Retired FDA Pharmacologists" (hereafter, "Retired FDA Pharmacologists"), (Rockville, Md.: National Library of Medicine, 1980), 10–16.

67. Ibid.

68. Ibid.

69. Chester I. Bliss, "The Calculation of the Dose-Mortality Curve," *Annals of Applied Biology* 22 (1935): 166.

228

NOTES TO PAGES 28–36

70. Edwin P. Laug et al., "The Toxicology of Some Glycols and Derivatives," *J. Industrial Hygiene and Toxicology* 21 (5) (1939): 200.

71. Ibid.

72. For examples of the development of chronic toxicity profiles for lead and other heavy metals, see Sellers, *Hazards of the Job*, 81–98, and Warren, *Brush with Death*, 64–83.

73. Vos et al., "Retired FDA Pharmacologists," 17–18.

74. Ibid., 24.

75. See Ross and Amter, *Polluters*.

76. See, for example, H. O. Calvery, E. P. Laug, and H. J. Morris, "The Chronic Effects on Dogs of Feeding Diets Containing Lead Acetate, Lead Arsenate, and Arsenic Trioxide in Varying Concentrations," *J.P.E.T.* 64 (4) (1938): 364–387, and Lucy L. Finner and H. O. Calvery, "Pathologic Changes in Rats and in Dogs Fed Diets Containing Lead and Arsenic Compounds," *Archives of Pathology* 27 (3) (1938): 433–466.

77. Bosso, *Pesticides and Politics*, 50.

78. Wallace, *Elixir Sulfanilamide*, 9.

79. Ibid.

80. In July 1934 Oettingen became the first director of Haskell Laboratory, an in-house medical research facility, at DuPont, one of the first such facilities in an American company. See Hounshell and Smith, *Science and Corporate Strategy*, 555–572.

81. W. F. von Oettingen and E. A. Jirouch, "The Pharmacology of Ethylene Glycol and Some of Its Derivatives in Relation to Chemical Constitution and Physical Chemical Properties," *J.P.E.T.* 42 (3) (1931): 371.

82. Young, "Sulfanilamide and Diethylene Glycol," 112.

83. Ibid.

84. Ibid. Estimates of the settlements varied from a confirmed amount of $2,000 to an FDA report of a rumor that S. E. Massengill paid out more than $500,000 in damage suit settlements.

85. Wallace, *Elixir Sulfanilamide*, 10.

86. Carpenter, *Reputation and Power*, 100.

87. E. M. K. Geiling, "Therapeutic Applications of Sulfanilamide and Allied Compounds," *Illinois Medical Journal* (November 1940): 404–405.

88. Ibid., 405.

89. Ibid.

90. Ibid., 410.

91. Cited in Anon., "Bristol Calls Ads Safety Insurance," *New York Times* (December 8, 1937), 38.

92. Ibid.

93. Morris Fishbein, "Elixir of Sulfanilamide Deaths and New Legislation," *Hygeia* 15 (1937): 1067.

94. Hilts, 89. Hilts's emphasis.

229

NOTES TO PAGES 37-42

95. This section is drawn from Young, "Sulfanilamide and Diethylene Glycol." See also Bosso, *Pesticides and Politics*, 51–52.

96. Gwen Kay, "Healthy Public Relations: The FDA's 1930s Legislative Campaign," *Bulletin of the History of Medicine* 75 (3) (Fall 2001): 446–487.

Chapter 2. DDT and Environmental Toxicology

1. See "War on Insects," *Time* 46 (August 27, 1945): 67. For a meticulous analysis of the development of DDT and its role in World War II, see Edmund Russell, *War and Nature: Fighting Humans and Insects with Chemicals from World War I to Silent Spring* (Cambridge: Cambridge University Press, 2001), and Russell, "War on Insects: Warfare, Insecticides, and Environmental Change in the United States, 1870–1945" (Ph.D. diss., University of Michigan, 1993). For another perspective on DDT and the war, see John H. Perkins, "Reshaping Technology in Wartime: The Effect of Military Goals on Entomological Research and Insect-Control Practices," *Technology and Culture* 19 (1978): 169–186. For more general historical analyses of insecticides, including DDT, see Thomas R. Dunlap, *DDT: Scientists, Citizens, and Public Policy* (Princeton: Princeton University Press, 1981); Whorton, *Before Silent Spring*; John H. Perkins, *Insects, Experts, and the Insecticide Crisis: The Quest for New Management Strategies* (New York: Plenum Press, 1982); Paul W. Riegert, *From Arsenic to DDT: A History of Entomology in Western Canada* (Toronto: University of Toronto Press, 1980); and David Kinkela, *DDT and the American Century: Global Health, Environmental Politics, and the Pesticide That Changed the World* (Chapel Hill: University of North Carolina Press), 2010.

2. For a thorough study of Müller and DDT's early history, see Russell, *War and Nature*.

3. Paul Herman Müller, *Histoire du DDT* (1948). Cited by Perkins, "Reshaping Technology in Wartime," 171.

4. See Perkins, "Reshaping Technology in Wartime," 171.

5. Ibid., 173.

6. For a thorough analysis of the evolution of the Bureau of Entomology and Plant Quarantine, see Russell, *War and Nature*, especially 264–348, and Perkins, "Reshaping Technology in Wartime," 169–186.

7. David Kinkela, *DDT and the American Century*, 16.

8. R. C. Bushland et al., "DDT for the Control of Human Lice," *J.E.E.* 37 (1) (1944): 126.

9. Kinkela, *DDT and the American Century*, 22.

10. See ibid., 35–61.

11. E. F. Knipling, "Insect Control Investigations of the Orlando, Fla., Laboratory during World War II," in *Annual Report of the Board of Regents of the Smithsonian Institution* (Washington, D.C.: Smithsonian Institution, 1948), 336.

12. DDT's efficacy was the subject of numerous reports in an issue of the *J.E.E.* in 1944.

13. A. H. Madden, A. W. Lindquist, and E. F. Knipling, "DDT as a Residual Spray for the Control of Bedbugs," *J.E.E.* 37 (1) (1944): 127.

14. H. K. Gouck and C. N. Smith, "DDT in the Control of Ticks on Dogs," *J.E.E.* 37 (1) (1944): 130.

15. J. B. Gahan and E. F. Knipling, "Efficacy of DDT as a Roach Poison," *J.E.E.* 37 (1) (1944): 139.

16. M. C. Swingle and E. L. Mayer, "Laboratory Tests of DDT against Various Insect Pests," *J.E.E.* 37 (1) (1944): 142.

17. O. A. Hills, "Tests with DDT against Pentatomids, Mirids, the Bollworm, and the Cotton Aphid," *J.E.E.* 37 (1) (1944): 142.

18. J. C. Clark, "Tests of DDT Dust against a Stinkbug and the Cotton Leafworm," *J.E.E.* 37 (1) (1944): 144, and George L. Smith, "Tests with DDT against the Boll Weevil," *J.E.E.* 37 (1) (1944): 144.

19. J. W. Ingram, "Tests of DDT Dust against the Sugarcane Borer, the Yellow Sugarcane Aphid, and the Argentine Ant," *J.E.E.* 37 (1) (1944): 145.

20. E. E. Ivy, "Tests with DDT on the More Important Cotton Insects," *J.E E.* 37 (1) (1944): 142.

21. W. E. Fleming and R. D. Chisholm, "DDT as a Protective Spray against the Japanese Beetle," *J.E E.* 37 (1) (1944): 155.

22. L. F. Steiner, C. H. Arnold, and S. A. Summerland, "Laboratory and Field Tests of DDT for Control of the Codling Moth," *J.E.E.* 37 (1) (1944): 157.

23. E. C. Holst, "DDT as a Stomach and Contact Poison for Honeybees," *J.E.E.* 37 (1) (1944): 159.

24. C. C. Plummer, "DDT and the Mexican Fruitfly," *J.E.E.* 37 (1) (1944): 158.

25. For a complete description of war time research at the Orlando laboratory, see Knipling, "Insect Control Investigations of the Orlando, Fla., Laboratory," 331–348.

26. R. L. Metcalf et al., "Observations on the Use of DDT for the Control of *Anopheles quadrimaculatus,*" *P.H.R.* 60 (27) (1945): 773.

27. Survivors had the capacity to develop resistance.

28. P. N. Annand, "How about DDT?" in *Address before the 41st Annual Convention, National Audubon Society, October 22, 1945, New York, New York* (Washington, D.C.: USDA Agricultural Research Administration, 1945), 5.

29. Ibid.

30. Ibid., 6. This statement anticipated the debate inspired by *Silent Spring.*

31. Ibid.

32. M. I. Smith and E. F. Stohlman, "The Pharmacologic Action of 2,2 bis (p-Chlorophenyl) 1,1,1 Trichlorethane and Its Estimation in the Tissues and Body Fluids," *P.H.R.* 59 (1944): 985.

33. Ibid.

34. Ibid.

NOTES TO PAGES 48–56

35. Ibid.

36. R. D. Lillie and M. I. Smith, "Pathology of Experimental Poisoning in Cats, Rabbits, and Rats with 2,2 bis-Parachlorphenyl-1,1,1 Trichlorethane," *P.H.R.* 59 (1944): 984.

37. P. A. Neal et al., "Toxicity and Potential Dangers of Aerosols, Mists, and Dusting Powders Containing DDT," *P.H.R.*, Supplement No. 177 (1944): 4.

38. Ibid., 7.

39. Ibid., 14.

40. Ibid., 22.

41. Ibid., 26–27.

42. A. A. Nelson et al., "Histopathological Changes Following Administration of DDT to Several Species of Animals," *P.H.R.* 59 (31) (1944): 1009–1020.

43. G. Woodard, A. A. Nelson, and H. O. Calvery, "Acute and Subacute Toxicity of DDT (2,2,-bis (p-Chlorophenyl)-1,1,1-Trichloroethane) to Laboratory Animals," *J.P.E.T.* 82 (1944): 153.

44. Ibid., 156.

45. Ibid., 157–158.

46. J. H. Draize, A. A. Nelson, and H. O. Calvery, "The Percutaneous Absorption of DDT (2,2-bis (p-Chlorophenyl) 1,1,1-Trichloroethane) in Laboratory Animals," *J.P.E.T.* 82 (1944): 161.

47. Ibid., 166.

48. R. J. Bing, B. McNamara, and F. H. Hopkins, "Studies on the Pharmacology of DDT (2,2 bis-Parachlorophenyl-1,1,1,Trichloroethane): The Chronic Toxicity of DDT in the Dog," *Bulletin of Johns Hopkins Hospital* 78 (1945): 310.

49. Edwin P. Laug, "A Biological Assay Method for Determining 2, 2 bis (p-Chlorophenyl)-1,1,1 Trichloroethane (DDT)," *J.P.E.T.* 86 (1946): 324.

50. Vos et al., "Retired FDA Pharmacologists," 48–49.

51. E. P. Laug and O. G. Fitzhugh, "2,2-bis (p-Chlorophenyl)-1,1,1-Trichloroethane (DDT) in the Tissues of the Rat Following Oral Ingestion for Periods of Six Months to Two Years," *J.P.E.T.* 87 (1946): 23.

52. See H. O. Calvery, E. P. Laug, and H. J. Morris, "The Chronic Effects on Dogs of Feeding Diets Containing Lead Acetate, Lead Arsenate, and Arsenic Trioxide in Varying Concentrations," *J.P.E.T.* 64 (4) (1938): 364–387, and E. P. Laug, and H. P. Morris, "The Effect of Lead on Rats Fed Diets Containing Lead Arsenate and Lead Acetate," *J.P.E.T.* 64 (4) (1938): 388–410.

53. H. S. Telford and J. E. Guthrie, "Transmission of the Toxicity of DDT through the Milk of White Rats and Goats," *Science* 102 (2660) (1945): 647.

54. See Nancy Langston, *Toxic Bodies: Hormone Disruptors and the Legacy of DES*, New Haven: Yale University Press, 2010.

55. Clarence Cottam and Elmer Higgins, "DDT: Its Effect on Fish and Wildlife," *U.S. Department of the Interior, Fish and Wildlife Service Circular* 11 (1946): 11.

56. Ibid.

57. Don R. Coburn and Ray Treichler, "Experiments on Toxicity of DDT to Wildlife," *J.W.M.* 10 (3) (1946): 208–210.

58. Cottam and Higgins, "DDT: Its Effect on Fish and Wildlife," 12.

59. Eugene W. Surber, "Effects of DDT on Fish," *J.W.M.* 10 (3) (1946): 187.

60. Ibid., 188.

61. Cottam and Higgins, "DDT: Its Effect on Fish and Wildlife," 1.

62. Ibid. Emphasis added.

63. Ibid.

64. Ibid., 4.

65. Robert E. Stewart et al., "Effects of DDT on Birds at the Patuxent Research Refuge," *J.W.M.* 10 (3) (1946): 201, and Cottam and Higgins, "DDT: Its Effect on Fish and Wildlife," 5.

66. R. T. Mitchell, "Effects of DDT Spray on Eggs and Nestlings of Birds," *J.W.M.* 10 (3) (1946): 194, and Cottam and Higgins, "DDT: Its Effect on Fish and Wildlife," 5.

67. Cottam and Higgins, "DDT: Its Effect on Fish and Wildlife," 6, and Surber, "Effects of DDT on Fish," 184.

68. Clarence M. Tarzwell, "Effects of DDT Mosquito Larviciding on Wildlife. Part 1, The Effects on Surface Organisms of the Routine Hand Application of DDT Larvicides for Mosquito Control," *P.H.R.* 62 (15) (1947): 525.

69. Ibid., 526.

70. Ibid.

71. Ibid., 528.

72. See Donald Worster, *Nature's Economy: A History of Ecological Ideas,* 2nd ed., ed. Donald Worster and Alfred Crosby (New York: Cambridge University Press, 1994), especially 291–315.

73. Tarzwell, "Effects of DDT Mosquito Larviciding on Wildlife," 530.

74. Ibid., 530–531.

75. Ibid., 545–546.

76. Ibid., 554.

77. Arnold B. Erickson, "Effects of DDT Mosquito Larviciding on Wildlife. Part 2, Effects of Routine Airplane Larviciding on Bird and Mammal Populations," *P.H.R.* 62 (1947): 1257.

78. Ibid., 1259.

79. Ibid., 1261.

80. P. A. Neal, W. F. Von Oettingen, and W. W. Smith, "Toxicity and Potential Dangers of Aerosols, Mists, and Dusting Powders Containing DDT," *P.H.R.* Supplement 177 (1944): 10.

81. Ibid., 12.

82. G. R. Cameron, "Risks to Man and Animals from the Use of 2,2-bis (p-Chlorophenyl), 1,1,1,-Trichlorethane (DDT): With a Note on the Toxicology of Gamma-Benzene Hexachloride (666, Gammexane)," *British Medical Bulletin* 3 (1945): 234.

NOTES TO PAGES 67-73

83. R. A. M. Case, "Toxic Effects of 2,2-bis (p-Chlorphenyl) 1,1,1-Trichlorethane (D.D.T.) in Man," *British Medical Journal* 2 (1945): 843.

84. Ibid.

85. F. M. G. Stammers and F. G. S. Whitfield, "The Toxicity of DDT to Man and Animals: A Report on the Work Carried Out at the Royal Naval School of Tropical Hygiene, Colombo, and a Review of the World Literature to January 1947," *Bulletin of Entomological Research* 38 (1) (1947): 580.

86. See P. A. Neal, T. R. Sweeney, S. S. Spicer, and W. F. von Oettingen, "The Excretion of DDT (2,2-Bis-(P-Chlorophenyl)-1,1,1-Trichloroethane) in Man, Together with Clinical Observations," *P.H.R.* 61: 403–409, and E.F. Stohlman and M. I. Smith, 1945, "The Isolation of Di(P-Chlorophenyl) Acetic Acid (DDA) from the Urine of Rabbits Poisoned with 2,2 Bis (P-Chlorophenyl) 1,1,1 Trichlorethane (DDT)," *J.P.E.T.* 84 (1946): 375–379.

87. E. P. Laug, F. M. Kunze, and C. S. Prickett, "Occurrence of DDT in Human Fat and Milk," *A.M.A Archives of Industrial Hygiene and Occupational Medicine* 3 (3) (1951): 245–246.

88. G. W. Pearce, A. W. Mattson, and W. J. Hayes, Jr., "Examination of Human Fat for the Presence of DDT," *Science* 116 (1952): 256.

89. Wayland J. Hayes, Jr., William F. Durham, and Cipriano Cueto, Jr., "The Effect of Known Repeated Oral Doses of Chlorophenothane (DDT) in Man," *J.A.M.A.* 162 (9) (1956): 891.

90. Ibid., 897.

91. Ibid.

92. See Robert L. Rudd and Richard E. Genelly, *Pesticides: Their Use and Toxicity in Relation to Wildlife* (Davis: California Department of Fish and Game, 1956); Dunlap, *DDT;* Dunlap, "Science as a Guide in Regulating Technology: The case of DDT in the United States," *Social Studies of Science* 8 (3) (1978): 265–285, and Robert N. Proctor, *Cancer Wars.*

93. Russell, "War on Insects," 447.

94. See Whorton, *Before Silent Spring.*

Chapter 3. The University of Chicago Toxicity Laboratory

1. See Daniel J. Kevles, *The Physicists: The History of a Scientific Community in Modern America* (New York: Alfred A. Knopf, 1978). For additional studies of contributions of the University of Chicago to the biological sciences, see several papers in Gregg Mitman, Jane Maienschein, and Adele E. Clarke, eds., "Crossing the Borderlands: Biology at Chicago," *Perspectives on Science: Historical, Philosophical, Social* 1 (3) (1993).

2. Biographical information on E. M. K. Geiling can be found in Philip C. Hoffmann and Alfred Heller, "E. M. K. Geiling (1891–1971)," in *Remembering the University of Chicago: Teachers, Scientists, and Scholars,* ed. Edward Shilts (Chicago: University of Chicago Press, 1991), 147–56.

234

NOTES TO PAGES 73–79

3. Linda Bren, "Frances Oldham Kelsey: FDA Medical Reviewer Leaves Her Mark on History," *FDA Consumer Magazine* (March–April 2001), http://web.archive.org/web/20061020043712/http://www.fda.gov/fdac/features/2001/201_kelsey.html. Accessed July 25, 2012.

4. John O. Hutchens, "The Tox Lab," *Scientific Monthly* 66 (1948): 107–108.

5. For a complete history of the OSRD, the NDRC, and CMR, see Irvin Stewart, *Organizing Scientific Research for War: The Administrative History of the Office of Scientific Research and Development* (Boston: Little, Brown, 1948). For a deeper analysis of the evolution of postwar science, see Roger L. Geiger, *Research and Relevant Knowledge* (New York: Oxford University Press, 1993), especially 3–29. For a detailed study of the Chemical Warfare Service and the OSRD, see Russell, *War and Nature*.

6. Oscar Bodansky, "Contributions of Medical Research in Chemical Warfare to Medicine," *Science* 102 (2656) (1945): 518. See also W. R. Kirner, "The Toxicity and Vesicancy of Chemical Warfare Agents," in *Chemistry: A History of the Chemistry Components of the National Defense Research Committee, 1940–1946*, ed. W. A. Noyes, Jr. (Boston: Little, Brown, 1948), 243–248.

7. John Doull, "Toxicology Comes of Age," *Annual Review of Pharmacology and Toxicology* 41 (2001): 2.

8. Hutchens, "Tox Lab," 108.

9. Ibid.

10. See Frederick Rowe Davis, "On the Professionalization of Toxicology," *Environmental History* 13 (October 2008): 751–756.

11. Doull, "Toxicology Comes of Age," 5.

12. Literally translated as "bad air," malaria had a long history in the warmer, tropical and subtropical regions of the world. See Paul F. Russell, *Man's Mastery of Malaria* (London: Oxford University Press, 1955). For detailed analysis, see Leo Barney Slater, *War and Disease: Biomedical Research on Malaria in the Twentieth Century*, New Brunswick, N.J.: Rutgers University Press, 2009.

13. William H. Taliaferro, "Malaria," in *Medicine and the War*, ed. Taliaferro (Chicago: University of Chicago Press, 1944), 55–75.

14. Ibid., 65.

15. Russell, *Man's Mastery of Malaria*, 108. See also Slater, *War and Disease*, pp. 59–108.

16. Graham Chen, "The Nature of the Enzyme Systems Present in *Trypanosoma equiperdum*: Introductory Statement," Grant Application, Department of Pharmacology, University of Chicago, n.d., 1–2.

17. Nathaniel Comfort, "The Prisoner as Model Organism: Malaria Research at Stateville Penitentiary," *Studies in History and Philosophy of Biological and Biomedical Sciences* 40 (2009): 192.

18. Chen, "Nature of the Enzyme Systems Present in *Trypanosoma equiperdum*," 4.

19. Ibid., 7–8.

235

NOTES TO PAGES 79–84

20. Ibid., 12–13.

21. F. E. Kelsey et al., "Studies on Antimalarial Drugs: The Excretion of Atabrine in the Urine of the Human Subject," *J.P.E.T.* 80 (1944b): 385.

22. See also Frederick Rowe Davis, "Unraveling the Complexities of Joint Toxicity of Multiple Chemicals at the Tox Lab and the FDA," *Environmental History* 13 (October 2008): 674–683.

23. Cited in C. I. Bliss, "The Toxicity of Poisons Applied Jointly," *Annals of Applied Biology* 26 (1939): 586.

24. Bliss, "Toxicity of Poisons," 585–587.

25. Graham Chen and E. M. K. Geiling, "The Acute Joint Toxicity of Atabrine, Quinine, Hydroxyethylapocupreine, Pamaquine and Pentaquine," *J.P.E.T.* 91 (1947): 138–139.

26. Ibid., 138.

27. Graham Chen and E. M. K. Geiling, "Trypanocidal Activity and Toxicity of Antimonials," *Journal of Infectious Disease* 76 (1945): 150.

28. Graham Chen and E. M. K. Geiling, "The Determination of Antitrypanosome Effect of Antimonials in Vitro," *Journal of Infectious Disease* 77 (1945): 142.

29. F. W. Schueler, G. Chen, and E. M. K. Geiling, "The Mechanism of Drug Resistance in Trypanosomes," *Journal of Infectious Disease* 81 (1947): 17.

30. Stanford Moore and W. R. Kirner, "The Physiological Mechanism of Action of Chemical Warfare Agents," in *Chemistry: A History of the Chemistry Components of the National Defense Research Committee, 1940–1946,* ed. W. A. Noyes, Jr. (Boston: Little, Brown, 1948), 249–260.

31. L. O. Jacobson, C. L. Spurr, E. S. G. Barron, T. Smith, C. Lushbaugh, and G. F. Dick, "Nitrogen Mustard Therapy: Studies on the Effect of Methyl-bis (Beta-chloroethyl) Amine Hydrochloride on Neoplastic Diseases and Allied Disorders of the Hemopoietic System," *J.A.M.A.* 132 (5) (1946): 263–71, and C. L. Spurr, L. O. Jacobson, T. R. Smith, and E. S. G. Barron, "The Clinical Application of a Nitrogen Mustard Compound Methyl bis (Beta-Chloroethyl) Amine to the Treatment of Neoplastic Disorders of the Hemopoietic System," *Cancer Research* 7 (1) (1947): 51–52.

32. Hoffmann and Heller "E. M. K. Geiling," 153.

33. See Gretchen Krueger, "The Formation of the American Society for Clinical Oncology and the Development of a Medical Specialty, 1964–1973," *Perspectives in Biology and Medicine* 47 (4) (Autumn 2004): 539. See also Krueger, *Hope and Suffering: Children, Cancer, and the Paradox of Experimental Medicine* (Baltimore: Johns Hopkins University Press, 2008).

34. Hutchens, "Tox Lab," 109–110.

35. E. M. K. Geiling, "Pharmacology," *Annual Review of Physiology* 10 (1948): 409–410.

NOTES TO PAGES 84–93

36. See Alice Kimball Smith, *A Peril and a Hope: The Scientists' Movement in America, 1945–47* (Chicago: University of Chicago Press, 1965), 539–559 and Angela N. H. Creager, *Life Atomic: A History of Radioisotopes in Science and Medicine* (Chicago: University of Chicago Press, 2013).

37. Eugene M. K. Geiling, "The Use of the Radioisotopes as an Experimental Tool," *Transactions and Studies of the College of Physicians of Philadelphia* 25 (1957): 57.

38. E. M. K. Geiling, B. J. McIntosh, and A. Ganz, "Biosynthesis of Radioactive Drugs Using Carbon 14," *Science* 108 (1948): 559.

39. E. M. K. Geiling et al., "Biosynthesis of Radioactive Medicinally Important Drugs with Special Reference to Digitoxin," *Transactions of the Association of American Physicians* 63 (1950): 94.

40. Ibid., 95.

41. John Doull, Kenneth P. DuBois, and E. M. K. Geiling, "The Biosynthesis of Radioactive Bufagin," *Archives of Internal Pharmacodynamics* 86 (4) (1951): 463.

42. George Okita, Robert B. Gordon, and E. M. K. Geiling, "Placental Transfer of Radioactive Digitoxin in Rats and Guinea Pigs," *Proceedings of the Society for Experimental Biology and Medicine* 80 (1952): 538.

43. G. T. Okita et al., "Studies on the Renal Excretion of Radioactive Digitoxin in Human Subjects with Cardiac Failure," *Circulation* 7 (1953): 167.

44. George T. Okita et al., "Blood Level Studies of C14-Digitoxin in Human Subjects with Cardiac Failure," *J.P.E.T.* 113 (1955): 380.

45. George T. Okita et al., "Metabolic Fate of Radioactive Digitoxin in Human Subjects," *J.P.E.T.* 115 (1955): 378.

46. John Doull, interview by author, October 27, 2000.

47. Ibid. See also Doull, "Toxicology Comes of Age," 5.

Chapter 4. The Toxicity of Organophosphate Chemicals

1. See Russell, *War and Nature*.

2. DuBois had also completed an M.S. in pharmaceutical chemistry at Purdue University and a B.S. in chemistry and pharmacy at South Dakota State University, where he participated in research on selenium poisoning as an undergraduate. For additional biographical information, see F. K. Kinoshita, "Kenneth Patrick DuBois (August 9, 1917—January 24, 1973)," *Toxicology and Applied Pharmacology* 25 (1973): 435–436, and John Doull, "Kenneth Patrick DuBois (August 9, 1917–January 24, 1973)," *Toxicological Sciences* 54 (2000): 1–2.

3. See Russell, *War and Nature*.

4. Hutchens, "Tox Lab," 111.

5. Doull, "Toxicology Comes of Age," 2–3.

NOTES TO PAGES 95–102

6. Kenneth P. DuBois and George H. Mangun, "Effect of Hexaethyl Tetraphosphate on Choline Esterase *in Vitro* and *in Vivo*," *Proceedings of the Society of Experimental Biology and Medicine* 64 (1947): 139.

7. Doull received his B.S. in chemistry at Montana State College in 1944, then spent two years in the navy as a radar and electronics specialist on the battleship *New Jersey* in the South Pacific. In 1946, he began a doctorate in biochemistry at the University of Chicago, on the advice of one of his professors at Montana State who arranged an interview with DuBois's colleague George Mangun (also a Montana graduate). It was Mangun who suggested that Doull switch to pharmacology to work with DuBois as his graduate advisor. See Doull, "Toxicology Comes of Age," 3.

8. Kenneth P. DuBois et al., "Studies on the Toxicity and Mechanism of Action of P-Nitrophenyl Diethyl Thionophosphate (Parathion)," *J.P.E.T.* 95 (1949): 79–91.

9. Cited in Kenneth P. DuBois, John Doull, and Julius M. Coon, "Studies on the Toxicity and Pharmacological Action of Octamethyl Pyrophosphoramide (OMPA; Pestox III)," *J.P.E.T.* 99 (1950): 376–393.

10. Ibid.

11. David Grob, William L. Garlick, and A. McGehee Harvey, "The Toxic Effects in Man of the Anticholinesterase Insecticide Parathion (P-Nitrophenyl Diethyl Thionophosphate)," *Johns Hopkins Hospital Bulletin* 87 (1950): 107.

12. Ibid., 127.

13. P. Lesley Bidstrup, "Poisoning by Organic Phosphorous Insecticides," *British Medical Journal* (1950): 548.

14. Ibid., 550.

15. Biographical material on Arnold J. Lehman is surprisingly limited given his long tenure at the FDA, but see Harry W. Hays, "Obituary: Arnold J. Lehman (September 2, 1900–July 10, 1979)," *Toxicology and Applied Pharmacology* 51 (1979): 549–551 and Anon., "About the Authors: Arnold J. Lehman, M.D.," *Food Drug Cosmetic Law Journal* 8 (7) (1953): 403.

16. Arnold J. Lehman, "The Toxicology of the Newer Agricultural Chemicals," *Quarterly Bulletin of the Association of Food and Drug Officials* 12 (3) (1948): 83–85.

17. Ibid., 87.

18. Ibid., 88.

19. Ibid.

20. Ibid., 68.

21. Ibid., 70.

22. Ibid.

23. Council on Pharmacy and Chemistry Committee on Pesticides, American Medical Society, "Pharmacology and Toxicology of Certain Organic Phosphorous Insecticides," *J.A.M.A.* 144 (2) (1950): 104–108.

238

NOTES TO PAGES 102–110

24. Ibid., 107–108.

25. Ibid., 108.

26. Ibid.

27. Ibid.

28. See Bosso, *Pesticides and Politics,* 66.

29. Kenneth P. DuBois, "Food Contamination from the New Insecticides," *Journal of the American Dietetic Association* 26 (1950): 326.

30. Ibid.

31. Ibid., 328.

32. Doull, "Toxicology Comes of Age," 3–4.

33. Kenneth P. DuBois and Julius M. Coon, "Toxicology of Organic Phosphorus-Containing Insecticides to Mammals," *A.M.A. Archives of Industrial Hygiene and Occupational Medicine* 6 (1952): 11–12.

34. Ibid., 12.

35. Lloyd W. Hazleton and Emily G. Holland, "Toxicity of Malathon: Summary of Mammalian Investigations," *A.M.A. Archives of Industrial Hygiene and Occupational Medicine* 8 (1953): 401. Emphasis added.

36. Ibid., 405.

37. Ibid.

38. Ibid.

39. Robert E. Bagdon and Kenneth P. DuBois, "Pharmacologic Effects of Chlorthion, Malathion and Tetrapropyl Dithionopyrophosphate in Mammals," *Archives of Internal Pharmacodynamics* 103 (2–3) (1955): 197.

40. C. P. Carpenter, H. F. Smyth, M. W. Woodside, P. E. Palm, C. S. Weil, and J. H. Nair, "Insecticide Toxicology—Mammalian Toxicity of 1-Naphthyl-n-Methylcarbamate (Sevin Insecticide)," *Journal of Agricultural and Food Chemistry* 9 (1) (1961): 30–39.

41. John E. Casida, Klas-Bertil Augustinsson, and Gunnel Jonsson, "Stability, Toxicity, and Reaction Mechanism with Esterases of Certain Carbamate Insecticides," *J.E.E.* 53 (2) (1960): 205–212. See also Robert Lee Metcalf, *Organic Insecticides, Their Chemistry and Mode of Action* (New York: Interscience Publishers), 1955.

42. Sheldon D. Murphy and Kenneth P. DuBois, "The Influence of Various Factors on the Enzymatic Conversion of Organic Thiophosphates to Anticholinesterase Agents," *J.P.E.T.* 124 (1958): 201.

43. Ibid.

44. Rachel Carson drew attention to endocrine system effects in *Silent Spring* in 1962. For a detailed history of endocrine disruptors, see Langston, *Toxic Bodies.* More than three decades later, a study by the National Research Council of the National Academy of Sciences of the basic variability in susceptibility to environmental and dietary chemicals between the young and the old led to the passage of the Food Quality Protection Act of 1996. For a thorough examination of the history, science, and policy leading to FQPA, see John Wargo, *Our Children's*

NOTES TO PAGES 111–119

Toxic Legacy: How Science and Law Fail to Protect Us from Pesticides, 2nd ed. (New Haven: Yale University Press, 1998).

45. J. William Cook, Fred L. Lofsvold, and James Harvey Young, "Interview between: J. William Cook, Retired Director, Division of Pesticide Chemistry and Toxicology, and Fred L. Lofsvold, FDA, and James Harvey Young, Emory University," in *History of the U.S. Food and Drug Administration* (Rockville, Md.: History of Medicine Division, National Library of Medicine, 1980), 1–2.

46. Ibid., 17–18.

47. Ibid., 18–19.

48. Ibid., 19–20.

49. Ibid., 21–22.

50. Ibid., 31–32, and J. William Cook, "In Vitro Destruction of Some Organophosphate Pesticides by Bovine Rumen Fluid," *Agricultural and Food Chemistry* 5 (11) (1957): 859–863.

51. D. F. McCaulley and J. William Cook, "A Fly Bioassay for the Determination of Organic Phosphate Pesticides," *Journal of the Association of Official Agricultural Chemists* 42 (1) (1959): 206.

52. John P. Frawley et al., "Marked Potentiation in Mammalian Toxicity from Simultaneous Administration of Two Anticholinesterase Compounds," *J.P.E.T.* 121 (1957): 96.

53. Ibid., 106.

54. Kenneth P. DuBois, "Potentiation of the Toxicity of Insecticidal Organic Phosphates," *A.M.A. Archives of Industrial Health* 18 (1958): 490–496.

55. Ibid., 495.

Chapter 5. What's the Risk?

1. S. R. Newell, *Federal Insecticide, Fungicide, and Rodenticide Act, Hearings Before the Committee on Agriculture House of Representatives on H.R. 4851 (H.R. 5645 reported):* "A bill to regulate the marketing of economic poisons and devices, and for other purposes," 79th Cong., February 5, 1946–April 11, 1947 (hereafter, *FIFRA Hearings*), 1.

2. Ibid., 4.

3. L. S. Hitchner, *FIFRA Hearings,* 29.

4. Ibid., 44.

5. Ibid., 46.

6. Ibid.

7. W. K. Granger, *FIFRA Hearings,* 46.

8. Hitchner, *FIFRA Hearings,* 46.

9. R. Smith, *FIFRA Hearings,* 84.

10. E. L. Griffin, *FIFRA Hearings,* 10.

240

NOTES TO PAGES 119–126

11. Bosso, *Pesticides and Politics,* 57.

12. Ibid., 57–58.

13. Whitaker, "Federal Pesticide Legislation in the U.S.," 449.

14. This section draws on Wargo, *Our Children's Toxic Legacy,* 70–71, and Bosso, *Pesticides and Politics,* 73–78.

15. Wargo, *Our Children's Toxic Legacy,* 71.

16. Bosso, *Pesticides and Politics,* 75.

17. K. T. Hutchinson, *Chemicals in Food Products: Hearings Before the House Select Committee to Investigate the Use of Chemicals in Food Products,* 82nd Cong., H. Res. 74 (hereafter, *Chemicals in Food Products Hearings*), September 14, 1950–March 6, 1952, 9.

18. R. L. Cleere, Letter to James J. Delaney, November 24, 1950, in *Chemicals in Food Products Hearings,* 39.

19. Ibid.

20. Carl E. Weigele, Letter to James J. Delaney, November 9, 1950, in *Chemicals in Food Products Hearings,* 53.

21. George A. Spendlove, Letter to James J. Delaney, October 27, 1950, in *Chemicals in Food Products Hearings,* 58.

22. Morton S. Biskind, *Chemicals in Food Products Hearings,* December 12, 1950, 700.

23. Ibid, 701.

24. Morton S. Biskind, "DDT Poisoning and Elusive Virus X: A New Cause for Gastro-enteritis," *American Journal of Digestive Diseases* 16 (March 1949): 79–84; Biskind, "Endocrine Disturbances in Gastrointestinal Conditions," *Review of Gastroenterology* 16 (March 1949): 220–225; and Biskind and Irving Bieber, "DDT Poisoning: A New Syndrome with Neuropsychiatric Manifestations," *American Journal of Psychotherapy* 3 (April 1949): 261–270.

25. A. I. Miller, *Chemicals in Food Products Hearings,* December 12, 1950, 705.

26. Ibid., 706.

27. E. M. Hedrick, *Chemicals in Food Products Hearings,* December 12, 1950, 707.

28. Biskind, *Chemicals in Food Products Hearings,* December 12, 1950, 707.

29. Ibid., 714.

30. Ibid., 716.

31. Ibid., 717.

32. Kleinfeld, *Chemicals in Food Products Hearings,* December 12, 1950, 717.

33. PHS and USDA, press release re: DDT, April 1, 1949, in *Chemicals in Food Products Hearings,* 719.

34. Wayland J. Hayes, Jr., *Chemicals in Food Products Hearings,* April 17, 1951, 90.

35. Ibid., 91.

241

NOTES TO PAGES 127–139

36. Ibid., 93.

37. Ibid., 96.

38. Ibid., 97.

39. Paul A. Neal, *Chemicals in Food Products Hearings,* April 17, 1951, 107.

40. Hayes, *Chemicals in Food Products Hearings,* 109.

41. Ibid., 111.

42. Frank Princi, *Chemicals in Food Products Hearings,* May 1, 1950, 149.

43. Ibid.

44. Miller, *Chemicals in Food Products Hearings,* 149.

45. Princi, *Chemicals in Food Products Hearings,* 150.

46. Ibid.

47. Ibid., 151.

48. Charles E. Palm, *Chemicals in Food Products Hearings,* May 1, 1950, 166.

49. Ibid., 167–168.

50. Ibid., 168.

51. Ibid.

52. Ibid., 169.

53. Ibid., 172.

54. George C. Decker, *Chemicals in Food Products Hearings,* May 2, 1951, 183.

55. Ibid.

56. Ibid.

57. John Dendy, *Chemicals in Food Products Hearings,* May 8, 1951, 220.

58. E. H. Hedrick, *Chemicals in Food Products Hearings,* 220.

59. Dendy, *Chemicals in Food Products Hearings,* 220.

60. Ibid., 222.

61. There is just one reference to Dendy unrelated to pesticides in Cyrus Longworth Lundell, *Agricultural Research at Renner, 1946–1967* (Renner, Tex.: Texas Research Foundation), 1967.

62. Dendy, *Chemicals in Food Products Hearings,* May 8, 1951, 237.

63. Louis Bromfield, *Chemicals in Food Products Hearings,* May 11, 1951, 292.

64. Ibid.

65. Ibid., 313.

66. Harold P. Morris, *Chemicals in Food Products Hearings,* May 11, 1951, 348.

67. Wilhelm C. Hueper, *Chemicals in Food Products Hearings,* January 29, 1952, 1370.

68. Ibid.

69. Ibid. Emphasis added.

70. Proctor, *Cancer Wars,* 47–48.

71. Ross and Amter, *Polluters,* 71.

72. Ibid.

73. See Sellers, "Discovering Environmental Cancer," 1832.

74. Hueper, *Chemicals in Food Products Hearings,* 1380.

NOTES TO PAGES 139-150

75. Ibid., 1381.

76. Fred C. Bishopp, *Chemicals in Food Products Hearings*, May 15, 1951, 373.

77. Ibid.

78. Ibid., 374.

79. Ibid., 375.

80. Ibid., 377. Emphasis added.

81. Ibid., 378. Emphasis added.

82. Ibid., 380.

83. Ibid.

84. Kleinfeld, *Chemicals in Food Products Hearings*, 520.

85. Bishopp, *Chemicals in Food Products Hearings*, 520.

86. Edward F. Knipling, *Chemicals in Food Products Hearings*, May 15, 1951, 521-522.

87. Quoted by Kleinfeld, *Chemicals in Food Products Hearings*, 523.

88. Miller, *Chemicals in Food Products Hearings*, 524.

89. Bishopp, *Chemicals in Food Products Hearings*, 524.

90. Knipling, *Chemicals in Food Products Hearings*, 525.

91. Ibid., 313.

92. Kleinfeld, *Chemicals in Food Products Hearings*, June 14, 1951, 518.

93. Bishopp, *Chemicals in Food Products Hearings*, 519. Emphasis added.

94. Arnold J. Lehman, *Chemicals in Food Products Hearings*, November 28, 1950, 389.

95. Ibid., 407.

96. Ibid.

97. Ibid., 408.

98. Ibid.

99. Kleinfeld, quoting Arnold Lehman, *Chemicals in Food Products Hearings*, 540.

100. Lehman, *Chemicals in Food Products Hearings*, 389

101. See Pete Daniel, *Toxic Drift: Pesticides and Health in the Post–World War II South* (Baton Rouge: Louisiana State University Press, 2005).

102. L. G. Cox, *Chemicals in Food Products Hearings*, January 31, 1952, 1388.

103. Ibid., 1390-1394.

104. Ibid., 1397.

105. For a similar conclusion, see Bosso, *Pesticides and Politics*, 75.

106. Ibid.

107. Wargo, *Our Children's Toxic Legacy*, 106.

108. Bosso, *Pesticides and Politics*, 77.

109. Wargo, *Our Children's Toxic Legacy*, 106.

110. Quoted in Bosso, *Pesticides and Politics*, 97.

111. Wargo, *Our Children's Toxic Legacy*, 107.

243

NOTES TO PAGES 150–159

112. Langston, *Toxic Bodies,* 82.

113. This section draws on Bosso, *Pesticides and Politics,* 98–100.

Chapter 6. Rereading *Silent Spring*

1. Doull, "Toxicology Comes of Age," 4.

2. Kenneth P. DuBois, "Proposed Research and Teaching Program in Toxicology at the University of Chicago," University Archives, University of Chicago, 1958, 1–2.

3. Doull, "Toxicology Comes of Age," 5.

4. Ibid.

5. Kenneth P. DuBois and E. M. K. Geiling, *Textbook of Toxicology* (New York: Oxford University Press, 1959), 211.

6. Ibid.

7. Ibid., 213.

8. The *Textbook of Toxicology* was the only one in its field until 1968, when Theodore Loomis and Wallace Hayes published a similar work. Another seven years passed before Doull, in collaboration with Louis Casarett, published *Toxicology: The Basic Science of Poisons* in 1975. In addition to providing the standard review of classes of toxic agents (metals, solvents, pesticides, etc.), this last text presented the organ system involved (kidney, liver, etc.) Over thirty-nine years and eight editions, *Casarett and Doull's Toxicology* (as the text became known) has remained the preferred textbook of toxicology. See John Doull, "Toxicology Comes of Age," 6–7. See also Mary O. Amdur, John Doull, and Curtis D. Klaassen, eds., *Casarett and Doull's Toxicology: The Basic Science of Poisons,* 4th ed. (New York: Pergamon Press, 1991).

9. Frederick Coulston, Arnold J. Lehman, and Harry W. Hays, Editors' Preface, *Toxicology and Applied Pharmacology* 1 (1959): iii.

10. Ibid., iii–iv.

11. Doull, "Toxicology Comes of Age," 5.

12. Martin W. Williams, Henry N. Fuyat, and O. Garth Fitzhugh, "The Subacute Toxicity of Four Organic Phosphates to Dogs," *Toxicology and Applied Pharmacology* 1 (1959): 1.

13. Frederick Coulston, Victor Drill, William Deichman, Harry Hays, Harold Hodge, Arnold Lehman, Boyd Schafer, Kenneth DuBois, and Paul Larson. See Harry W. Hays, *Society of Toxicology History, 1961–1986* (Washington, D.C.: Society of Toxicology, 1986).

14. Doull, "Toxicology Comes of Age," 7.

15. Ibid., 7–8. Emphasis added.

16. See Robert Rudd, *Pesticides and the Living Landscape* (Madison: University of Wisconsin Press, 1964), and Lewis Herber, *Our Synthetic Environment* (New York: Alfred A, Knopf, 1962).

17. Rachel Carson, *Silent Spring* (Boston: Houghton Mifflin, 1962), 26–27.

244

NOTES TO PAGES 159–165

18. Ibid., 27–28.

19. Ibid., 29.

20. Ibid., 30.

21. Ibid., 31.

22. Ibid., 126–127.

23. Ibid., 196–198.

24. Ibid., 296.

25. Ibid., 297.

26. See, for example, Linda Lear, *Rachel Carson: Witness for Nature* (New York: Owl Books, 1998); Mark Lytle, *The Gentle Subversive: Rachel Carson, Silent Spring, and the Rise of the Environmental Movement* (Oxford: Oxford University Press, 2007); William Souder, *On a Farther Shore: The Life and Legacy of Rachel Carson* (New York: Crown Publishers, 2012); Dunlap, *DDT;* Maril Hazlett, "The Story of Silent Spring and the Ecological Turn," (Ph.D. diss., University of Kansas, 2003); Hazlett, " 'Woman vs. Man vs. Bugs:' Gender and Popular Ecology in Early Reactions to *Silent Spring*," *Environmental History* 9 (4) (October 2004): 701–729; Frank Graham, *Since Silent Spring* (Boston: Houghton-Mifflin, 1970); and especially Thomas R. Dunlap, *DDT, Silent Spring, and the Rise of Environmentalism: Classic Texts* (Seattle: University of Washington Press, 2008).

27. Joseph Hickey, quoted in "Interview with Joseph J. Hickey," Dunlap, *DDT,* 82.

28. *CBS Reports,* "The Silent Spring of Rachel Carson," April 3, 1963, transcript. Quoted in Lytle, *Gentle Subversive,* 183.

29. Frank Graham, Jr., *Since Silent Spring* (New York: Fawcett Crest, 1970), 61.

30. For a detailed analysis of the PSAC and its significance in cold war America, see Zuoyue Wang, *In Sputnik's Shadow: The President's Science Advisory Committee and Cold War America* (New Brunswick, N.J.: Rutgers University Press, 2008).

31. President's Science Advisory Committee, "Use of Pesticides: A Report by the President's Science Advisory Committee" (Washington, D.C.: The White House, May 15, 1963) (hereafter, PSAC, "Use of Pesticides"), 1.

32. Ibid.

33. Ibid., 3.

34. Ibid., 4. Emphasis added.

35. See Sarah A. Vogel, "From 'The Dose Makes the Poison' to 'The Timing Makes the Poison': Conceptualizing Risk in the Synthetic Age," *Environmental History* 13 (October 2008): 667–673, and Vogel, *Is It Safe? BPA and the Struggle to Define the Safety of Chemicals* (Berkeley: University of California Press, 2013). See also Proctor, *The Cancer Wars,* and Wargo, *Our Children's Toxic Legacy.*

36. PSAC, "Use of Pesticides," 13.

37. Ibid.

38. Ibid., 22.

NOTES TO PAGES 166–174

39. Abraham Ribicoff, *Interagency Coordination in Environmental Hazards (Pesticides), Hearings Before the Subcommittee on Reorganization and International Organizations of the Committee on Government Operations United States Senate*, 88th Cong., 1st Sess.), May 16, 1963–July 29, 1964 (hereafter, *Interagency Coordination in Environmental Hazards*), 1.

40. Ibid., 2.

41. Orville L. Freeman, *Interagency Coordination in Environmental Hazards*, 84.

42. Ibid., 87.

43. Ibid., 86.

44. Ibid., 98.

45. Ribicoff, *Interagency Coordination in Environmental Hazards*, 206.

46. Ibid.

47. Rachel Carson, *Interagency Coordination in Environmental Hazards*, 215–216.

48. Ibid., 217.

49. Ibid., 218–219.

50. Ribicoff, *Interagency Coordination in Environmental Hazards*, 220.

51. Carson, *Interagency Coordination in Environmental Hazards*, 244.

52. Ibid.

53. See Scott Frickel, *Chemical Consequences: Environmental Mutagens, Scientist Activism, and the Rise of Genetic Toxicology* (New Brunswick, N.J.: Rutgers University Press, 2004).

54. Carson., *Interagency Coordination in Environmental Hazards*, 246.

55. James B. Pearson, *Interagency Coordination in Environmental Hazards*, 247.

56. Ribicoff, *Interagency Coordination in Environmental Hazards*, 2005–2006.

57. Ibid., 2006.

58. PSAC, "Use of Pesticides," 13.

59. Ibid., 20.

60. George Larrick, *Interagency Coordination in Environmental Hazards*, 191.

61. Theron G. Randolph, *Interagency Coordination in Environmental Hazards*, 591.

62. Irma West, *Interagency Coordination in Environmental Hazards*, 601.

63. Ibid., 617.

64. Ibid., 626.

65. Julius E. Johnson, *Interagency Coordination in Environmental Hazards*, 1334.

66. Thomas H. Milby and Fred Ottoboni, Report of an Epidemic of Organic Phosphate Poisoning in Peach Pickers, Stanislaus County, California, August 1963. Exhibit 132 in *Interagency Coordination in Environmental Hazards*, 1452.

NOTES TO PAGES 174–182

67. Bert J. Vos, *Interagency Coordination in Environmental Hazards*, 759.

68. Ribicoff, *Interagency Coordination in Environmental Hazards*, 760.

69. Vos, *Interagency Coordination in Environmental Hazards*, 760.

70. Recall that Frawley published several important papers on the toxicity of pesticides; see references in chapter 2.

71. Julius E. Johnson, *Interagency Coordination in Environmental Hazards*, 1335.

72. This section draws on Langston, *Toxic Bodies*, 90–95.

73. Roger Williams, "The Nazis and Thalidomide: The Worst Drug Scandal of All Time," *Newsweek* (September 10, 2012). Somewhat ironically this important piece of investigative journalism did not follow thalidomide to the United States.

74. Langston, *Toxic Bodies*, 93.

75. Ibid., 95.

76. Exhibit 126, *Interagency Coordination in Environmental Hazards*, 1342.

77. Ibid.

78. E. F. Feichtmeir, *Interagency Coordination in Environmental Hazards*, 1343.

79. See J. E. Lovelock, "A Sensitive Detector for Gas Chromatography," *Journal of Chromatography* 1 (1958): 35–46. See also John and Mary Gribbin, *James Lovelock: In Search of Gaia* (Princeton: Princeton University Press, 2009), 97–98.

80. Ribicoff, *Interagency Coordination in Environmental Hazards*, 1343.

81. For a detailed analysis of the precautionary principle, see Langston, *Toxic Bodies*, 152–166. See also Wargo, *Our Children's Toxic Legacy*.

82. Feichtmeir, *Interagency Coordination in Environmental Hazards*, 1344.

83. Ernest J. Jaworski, *Interagency Coordination in Environmental Hazards*, 1348.

84. Ibid.

85. See Bruce N. Ames, Margie Profet, and Lois Swirsky Gold, "Nature's Chemicals and Synthetic Chemicals: Comparative Toxicology," *Proceedings of the National Academy of Sciences* 87 (October 1990): 7782–7786.

86. Arnold Lehman, *Interagency Coordination in Environmental Hazards*, 1117.

87. Ibid., 1145.

88. Kenneth DuBois, *Interagency Coordination in Environmental Hazards*, 1246.

89. Ibid.

90. Ibid., 1247.

91. John P. Frawley, *Interagency Coordination in Environmental Hazards*, 1317.

92. Julius E. Johnson, *Interagency Coordination in Environmental Hazards*, 1409.

NOTES TO PAGES 183–192

93. Ibid., 1412.

94. Ibid., 1413.

95. Ibid.

96. Ernest G. Jaworski, *Interagency Coordination in Environmental Hazards*, 1429.

97. Ribicoff, *Interagency Coordination in Environmental Hazards*, 1342.

98. See Daniel, *Toxic Drift*, Langston, *Toxic Bodies*, Gerald Markowitz and David Rosner, *Deceit and Denial: The Deadly Politics of Industrial Pollution* (Berkeley: University of California Press, 2002), and Naomi Oreskes and Erik M. Conway, *Merchants of Doubt: How a Handful of Scientists Obscured the Truth on Issues from Tobacco Smoke to Global Warming* (New York: Bloomsbury Press, 2010). For an ethical perspective, see Kristin Shrader-Frechette, *Taking Action, Saving Lives: Our Duties to Protect Environmental and Public Health* (Oxford: Oxford University Press), 2007.

99. Daniel, *Toxic Drift*, 67–68.

100. Ibid., 83.

Chapter 7. Pesticides and Toxicology after the DDT Ban

1. Dunlap, *DDT.*

2. This section is drawn from Dunlap, *DDT,* 129–196.

3. J. Brooks Flippen, "Pests, Pollution, and Politics: The Nixon Administration's Pesticide Policy," *Agricultural History* 71 (4): 452.

4. Ibid., 197–245.

5. Anon., "Low Consumption of Insecticides Aids Resistance," *Medical Tribune* (October 5, 1964), 15.

6. Anon., "Pesticide Poisonings Widespread in U.S.," *Santa Ana Register* (September 22, 1969), 10.

7. Chicago Daily News Service, "Pesticide Poisoning Expected to Increase," *Springfield Daily News* (December 21, 1971), 1.

8. Ibid.

9. William D. Ruckleshaus, "Federal Register," (Washington, D.C.: 1972), cited in Lewis Regenstein, *America the Poisoned: How Deadly Chemicals Are Destroying Our Environment, Our Wildlife, Ourselves and—How We Can Survive!* (Washington, D.C.: Acropolis Books, 1982), 107.

10. This section draws on Flippen, "Pests, Pollution, and Politics," and Flippen, *Nixon and the Environment* (Albuquerque: University of New Mexico Press), 2000. See also Wargo, *Our Children's Toxic Legacy,* 89–93, and Mary Jane Large, "Comments: The Federal Pesticide Control Act of 1972: A Compromise Approach," *Ecology Law Quarterly* 3 (1973): 277–310.

11. Flippen, "Pests, Pollution, and Politics," 451–52.

12. Wargo, *Our Children's Toxic Legacy,* 89.

NOTES TO PAGES 192–209

13. Ibid., 90.

14. Ibid., 91.

15. Ibid.

16. Ibid., 92.

17. Ibid., 93.

18. Ibid., 93–94.

19. Ibid., 87–88.

20. Michael Elliott, "Properties and Applications of Pyrethroids," *E.H.P.* 14 (1976): 4.

21. John E. Casida, "Michael Elliott's Billion Dollar Crystals and Other Discoveries in Insecticidal Chemistry," *Pesticide Management Science* 66 (2010): 1163.

22. M. Elliot, A. W. Farnham, N. F. Janes, P. H. Needham, D. A. Pulman, and J. H. Stevenson, "A Photostable Pyrethroid," *Nature* 246 (November 16, 1973): 169.

23. Elliott, "Properties and Applications," 6.

24. Ibid., 9.

25. Ibid., 10

26. Ibid., 11.

27. John E. Casida, "Pyrethrum Flowers and Pyrethroid Insecticides," *E.H.P.* 34 (February 1984): 199.

28. Michael Elliott, "Progress in the Design of Insecticides," *Chemistry and Industry* (November 17, 1978): 759.

29. Ibid.

30. Ibid., 767.

31. Ibid.

32. Ibid., 767–768.

33. Leonard P. Gianessi, "U.S. Pesticide Use Trends, 1966–1989" (Washington, D.C.: Resources for the Future, 1992).

34. Ibid., 12.

35. Ibid., 13.

36. Ibid., 9.

37. "Chlorpyrifos," Pesticide Information Profile, Extension Toxicology Network. Available from http://pmep.cce.cornell.edu/profiles/extoxnet/carbaryl-dicrotophos/chlorpyrifos-ext.html. Accessed May 20, 2011.

38. Gianessi, "U.S. Pesticide Use Trends," 10.

39. *J.A.M.A.* 260 (7): 963.

40. Theo Colborn, "Pesticides: How Research Has Succeeded and Failed to Translate Science into Policy: Endocrinological Effects on Wildlife," *E.H.P.* 103, Supplement 6 (September 1995): 81–85. See also Colborn, Dianne Dumanoski, and John Peterson Myers, *Our Stolen Future: Are We Threatening Our Fertility, Intelligence, and Survival? A Scientific Detective Story* (New York: Dutton, 1996).

249

NOTES TO PAGES 209–212

41. See Langston, *Toxic Bodies.*

42. Wargo, *Our Children's Toxic Legacy,* 94.

43. Joel Bourne, "Buggin Out: Integrated Pest Management Uses Natural Solutions Both Old and New to Help Farmers Kick the Chemical Habit," *Audubon* 101 (2) (1999): 73.

44. American Bird Conservancy, *Monocrotophos.* Available from http://www.abcbirds.org/pesticides/Profiles/monocrotophos.htm (accessed March 2001).

45. American Bird Conservancy, *Diazinon.* Available from http://www.abc birds.org/pesticides/Profiles/diazinon.htm (accessed March 2001).

46. Ted Schettler et al., *Generations at Risk: Reproductive Health and the Environment* (Cambridge: MIT Press, 1999), 126–129.

47. Brenda Eskenazi, Asa Bradman, and Rosemary Castorina, "Exposures of Children to Organophosphate Pesticides and Their Potential Adverse Health Effects," *E.H.P.* 107, Supplement 3 (1999): 409.

48. Gardiner Harris and Hari Kumar, "Contaminated Lunches Kill 22 Children in India," *New York Times* (July 17, 2013), http://www.nytimes.com/2013/07/18/world/asia/children-die-from-tainted-lunches-at-indian-school.html. Accessed July 18, 2013. See also Meera Subramanian, "Bihar School Deaths Highlight India's Struggle with Pesticides," *New York Times* (July 30, 2013), http://india.blogs.nytimes.com/2013/07/30/bihar-school-deaths-highlight -indias-struggle-with-pesticides/. Accessed August 20, 2013.

49. Office of Pesticide Programs (USEPA), "Organophosphate Pesticides in Food: A Primer on Reassessment of Residue Limits," (Washington, D.C.: Environmental Protection Agency, 1999), 1.

50. Consumers Union, "Consumers Union Praises EPA Phase Out of the Pesticide Diazinon: Hopes New Administration Will Continue Rigorous Examination of Pesticides," Press Release, December 5, 2000, 1.

51. Virginia Rauh, Srikesh Arunajadai, Megan Horton, Frederica Perera, Lori Hoepner, Dana B. Barr, and Robin Whyatt, "Seven-Year Neurodevelopmental Scores and Prenatal Exposure to Chlorpyrifos, a Common Agricultural Pesticide," *E.H.P.* 119 (2011): 1196–1201. See also Theo Colborn, "A Case for Revisiting the Safety of Pesticides: A Closer Look at Neurodevelopment," *E.H.P.* 114 (1) (2006): 10–17.

52. Stephanie M. Engel, James Wetmur, Jia Chen, Chenbo Zhu, Dana Boyd Barr, Richard L. Canfield, and Mary S. Wolff, "Prenatal Exposure to Organophosphates, Paraoxonase 1, and Cognitive Development in Childhood," *E.H.P.* 119 (2011): 1182–1188.

53. Maryse F. Bouchard, Jonathan Chevrier, Kim G. Harley, Katherine Kogut, Michelle Vedar, Norma Calderon, Celina Trujillo, Caroline Johnson, Asa Bradman, Dana Boyd Barr, and Brenda Eskenazi, "Prenatal Exposure to Organophosphate Pesticides and IQ in 7-Year-old Children," *E.H.P.* 119 (2011): 1189–1195.

NOTES TO PAGES 212-222

54. John D. Meeker and Heather M. Stapleton, "House Dust Concentrations of Organophosphate Flame Retardants in Relation to Hormone Levels and Semen Quality Parameters," *E.H.P.* 119 (2010): 318–323.

Epilogue

1. Izuru Yamamoto, "Nicotine to Nicitinoids: 1962 to 1997," in Yamamoto and John Casida, *Nicotinoid Insecticides and the Nicotinic Acetylcholine Receptor* (Tokyo: Springer-Verlag, 2008), 3–27.

2. Motohiro Tomizawa and John E. Casida, "Neonicotinoid Insecticide Toxicology: Mechanisms of Selective Action," *Annual Review of Pharmacology and Toxicology* 45 (2005): 252.

3. Pierre Mineau and Cynthia Palmer, *The Impact of the Nation's Most Widely Used Insecticides on Birds* (Washington, D.C.: American Bird Conservancy, 2013), 5–9.

4. Dave Goulson, "An Overview of the Environmental Risks Posed by Neonicotinoid Insecticides," *Journal of Applied Ecology* 50 (4) (2013): 1–11.

5. See Mineau and Palmer, *Impact of the Nation's Most Widely Used Insecticides on Birds*, and Goulson, "An Overview of Environmental Risks Posed by Neonicitinoid Insecticides."

6. EPA, "Colony Collapse Disorder: European Bans on Neonicotinoid Pesticides," http://www.epa.gov/pesticides/about/intheworks/ccd-european-ban.html. Accessed September 13, 2013.

251

INDEX

accumulation. *See* bioaccumulation
acute vs. chronic toxicity: and cancer, 9,
    26, 127; and DDT analysis, 47–48, 49,
    51–55, 65, 69–70, 100, 127, 129, 132–133;
    and diethylene glycol, 24; and food
    residues, 12, 13, 69, 103–104, 178;
    importance of both, 28, 164; and joint
    toxicity, 81–82; and new-drug analysis,
    33; and organophosphates vs. chlori-
    nated hydrocarbons, 103–104, 115,
    140–141, 148–149, 151, 160–161, 164–165,
    191, 212; Public Health Service and, 29.
    See also specific insecticides
advertising, 9–10, 34–35, 170
Agricultural Insecticide and Fungicide
    Association, 116, 117–119
agriculture: arsenate use 1919–1929,
    10–11; importance of insecticides
    to, xiii, 1–2, 3–4, 122, 132, 141–142,
    163–164, 166–167; industrialized,
    xiii, 1–2, 3–4; insecticide use
    1966–1989, 203
aldicarb, 205, 206, 207, 208
aldrin, 158, 164, 189, 194, 200, 201, 202,
    204
alkyl pyrophosphates, 105
alkyl thiophosphates, 105–106

allethrin, 195, 196
AMA. *See* American Medical Association
    (AMA)
American Bird Conservancy, 222
*American Chamber of Horrors: The Truth
    about Food and Drugs* (Lamb), 10
American Chemical Society, 12, 144,
    189–190
American Cyanamid, 102–103, 106–108,
    115, 163
American Medical Association (AMA):
    and Elixir Sulfanilamide tragedy,
    20–23; on organophosphates,
    102–103; on pesticide-cancer link,
    124, 206, 209
Ames, Bruce, 179
aminotriazole, 150–151
amphibians, 57, 58, 59, 60
Amter, Steven, 138
Andresen, August, 119
Annand, Percy Nichol, 40, 45–46
aphids, 42, 43, 93–94, 135, 201
aquatic insects, 45, 61–64, 198
arsenates, 10–13. *See also* lead arsenate
Atomic Energy Commission, 84
Aub, Joseph, 7
Audubon, 210

253

INDEX

Bagdon, Robert, 108–109
bedbugs, 42
Beech-Nut Packing Company, 146–148
Before Silent Spring (Whorton), 11
BEPQ. *See* Bureau of Entomology and
    Plant Quarantine (BEPQ)
BHC (benzene hexachloride), 125, 140,
    142, 147
Bidstrup, Lesley, 98
bioaccumulation/accumulation: Carson
    on, 158–159, 187, 209; of DDT, 46,
    53–54, 55, 68–69, 100, 103–104, 134,
    137–138, 139, 141; of neonicitinoids, 222;
    of organophosphates, 97, 187, 217–218;
    of pyrethroids, 20
biological control, 161–162, 200
biomagnification, 134, 140–141, 165, 187
bioresmethrin, 196–197, 198
birds, ix–x, 56–57, 57–58, 59–60, 64, 198,
    210, 222
Bishopp, Fred C., 139–145, 146
Biskind, Morton S., 123–126, 127
Bliss, Chester I., 26–27, 80–81
bobwhite quail, 56–57
Bosso, Christopher, 25, 29, 121–122, 149
Braun, Herbert, 25
Bristol, Lee H., 34–35
Bromfield, Louis, 135–136
bromination technique (brominated spot
    test), 111–112
bufagin, 87
Bureau of Entomology and Plant
    Quarantine (BEPQ, USDA), 40–44,
    139–145
Bushland, Raymond C., 41

California Department of Health, 102
California Department of Public Health,
    172–173
Calvery, Herbert O., 26, 29, 51–52
Campbell, Walter, 9, 12–13, 25, 33
cancellation of registration for a
    pesticide, 191–192, 194
cancer: chemotherapy, 83; Delaney
    Clause and food residues and, 150;
    occupational, 8–9; and pesticides, 124,
    136–139, 206, 209

Cannan, Keith, 75
Cannon, Clarence, 103, 138
Cannon, Paul R., 24–25, 32
carbamates, 109, 196, 197, 198, 200–201, 211
carbaryl (Sevin), 109, 201, 202, 203, 207,
    208
carbofuran, 202, 203, 207, 208
Carpenter, Daniel, 33
Carson, Rachel, 157–158, 162–163, 168–171.
    See also *Silent Spring* (Carson)
Carter Memorial Laboratory, 61
Casida, John E., 195
CBS Reports, 162–163
Celluloid Corporation, 14–15
Center for Biological Diversity, x
central nervous system effects. *See*
    neurotoxicity/central nervous system
    effects
Chemagro, 94, 105
chemical industry (insecticide industry,
    pesticide industry): advertising, 9–10,
    34–35, 170; "capture" of USDA and
    FDA, 146, 185; and Delaney Hearings,
    121–122; and labeling, 5; on new-
    pesticide development and testing,
    182–184; on no-effect level, 174–178;
    and registration (at FIFRA hearings),
    117–119; response to *Silent Spring*, xii,
    162; support for Federal Environmen-
    tal Pesticide Control Act (FEPCA),
    192, 193. *See also specific companies*
Chemicals in Food Products Hearings.
    See Delaney Hearings (Chemicals in
    Food Products Hearings)
chemical warfare agents, 74–76
Chemical Warfare Service (CWS), 73–76,
    84
Chemical Week Reports, 184
Chemie Grünenthal, 175
Chen, Graham, 78–83
Chicago Dietetic Association, 103–104
Chicago stockyards, 1
children, 172, 174–176, 211, 212
chlordane, 125, 142, 146, 147, 204–205
chlorinated hydrocarbons: acute
    poisoning from, 190–191; banning of,
    xiv, 188–190, 194–195; and cancer,

254

INDEX

137–138, 194; Carson on, 157–159, 160–161; food residues and chronic toxicity, 100, 103–104, 133–134; and organophosphate trade-off, 140–141, 185–186, 187; resistance to, 131–132, 148, 160–161; usage 1966–1989, 200–206. *See also specific chlorinated hydrocarbons*

chloroquine, 77, 78

chlorpyrifos, 203, 205, 206, 207, 208, 211, 212

cholinesterase inhibition: carbamates and, 109; Carson on, 159; Cook's testing of organophosphates, 110–113; DuBois' research on organophosphates, 92–96, 191; Grob's research on organophosphates, 96–98; malathion and, 107; OMPA and, 106; Wargo on organophosphates and, 211

chromatography, 111–112

chromium dust research, 138–139

chronic toxicity. *See* acute vs. chronic toxicity; *specific insecticides*

Cleere, R. L., 122–123

Clement, Roland, 162

clothianidin, 221, 222

Coburn, Don R., 56–57

cockroaches, German, 42, 94, 197

Coggeshall, Lowell, 78, 79

Colborn, Theo, 209

Colorado Department of Health, 122–123

Comfort, Nathaniel, 78

Consumers' Research, Inc., 10

Conway, Eric M., 184–185

Cook, J. William, 110–113

Coon, Julius M., 95, 105–106, 107–108

Copeland, Royal S., 13, 36

corn, 202, 203

Cottam, Clarence, 58–59

cotton pests, 42–43, 202, 203

cottontail rabbits, 56

Coulston, Frederick, 155

Council on Scientific Affairs (CSA) of the American Medical Association, 206, 209

Cox, L. G., 146–148

cranberry scare, 150–151

cross-resistance, 198

Daniel, Pete, 146, 184–185

DDT: acute and chronic toxicity studies on laboratory animals, 47–55; acute poisonings and deaths, 47, 124, 127, 132–133; aerosol experiments, 48–50; application rates, 46, 141; banning of, ix, 187–189, 218–220; Beech-Nut Packing Company (Cox) on residues in baby food, 146–148; bioaccumulation, 46, 53–54, 55, 68–69, 100, 103–104, 134, 137–138, 139, 141; Biskind (physician) on clinical experiences with DDT, 123–126; Bromfield (farmer) on use of, 135–136; and cancer, 137–139; central nervous system effects, 48, 53, 65, 67; compared to organophosphates, 99–101; Delaney committee concerns about, 121; dermal absorption studies, 48, 66–67; and disease-carrying and pest insects, 40–41, 42, 44–45, 142; and Dutch elm disease control, 188; FDA Division of Pharmacology analysis, 50–54; Fish and Wildlife Service experiments, 55–60; food and milk residues and chronic toxicity, 54–55, 103–104, 123–125, 126–129, 133–135; and growth rates, 52; human experiments, 65–69; importance of, to agriculture and the food supply, 141–142; increased populations of nontarget insects after use of, 43; ingestion studies, 47–48, 51–52; initial synthesis of, 39; joint government agency press release on, 126; lab *vs.* field studies, 60–61, 62; liver pathology and, 48, 51; Müller's research at Geigy, 39–40; NIH analysis, 47–50; NIH inhalation experiments, 65–66; persistence of, 41, 140–141; PHS (Hayes and Neal) on residues in food, 126–129; PHS Savannah River National Wildlife Refuge DDT larviciding studies, 61–65; resistance to (acquired), 44, 131–133; resistant insects, 42–43; Tennessee Valley Authority (TVA) research, 44–45; Texas Research Foundation (Dendy) on food contamination

255

INDEX

DDT (continued)
and accumulation of, 133–135; USDA
Bureau of Entomology and Plant
Quarantine (BEPQ) analysis, 40–44,
45–46; USDA (Bishopp) on risks of,
139–145; wild animal laboratory
experiments, 55–58; wildlife field
studies, 58–65
Decker, George C., 132–133
Delaney, James J., 121
Delaney Clause (1958), 150–151
Delaney Hearings (Chemicals in Food
Products Hearings), 121–151; Bishopp
(USDA Bureau of Entomology and
Plant Quarantine) on risks of DDT,
139–145; Bishopp (USDA) on use and
toxicity of parathion, 144–145; Biskind's
clinical experiences with DDT, 123–126;
Bromfield's testimony on overuse of
insecticides, 135–136; carcinogenic
properties of insecticides, 136–139;
congressmen's concerns about DDT,
121; Cox's (Beech-Nut Packing
Company) testimony on residues in
baby food, 146–148; Dendy's testimony
on food contamination and accumulation of DDT (Texas Research Foundation), 133–135; economic entomologists
testimony on insects developing
resistance to DDT, 131–133; Hayes and
Neal's statements on DDT residues in
food (PHS), 126–129; joint government
agency press release on safety of DDT
(April 1, 1949), 126; and lack of coordination between agencies/professions,
134–135; Lehman (FDA) on chlordane
toxicity, 146; Lehman (FDA) on risks of
parathion, 145–146; overview, 151–152;
results of, 149–152; and scientific
uncertainty, 121, 129–133, 148–149; state
health officials concerns, 122–123;
testimony on importance of insecticides
to agriculture and the food supply, 122,
132, 141–142
Dendy, John, 133–135
dermal absorption, 48, 52, 66–67, 96, 99,
100

DFP (diisopropyl fluorophosphate), 92, 94
diazinon, 200, 201, 210, 211
dieldrin, 188, 189, 194
diethylene glycol, 19, 23–24, 30–32
digitoxin, 85–89
diisopropyl fluorophosphate (DFP), 92, 94
disulfoton, 201, 202
Domagk, Gerhardt, 18
dosage-mortality curve, 26–29
Doull, John, 90; on FDA and Chemagro
meetings, 105; on Geiling and Society
of Toxicology, 156; on Geiling's
leadership, 76; organophosphate
research, 95–96; and radioactive
bufagin, 87; and screening program for
radioactive elements, 89; toxicity data
base, 153–154; on *Toxicology and
Applied Pharmacology*, 155–156
Dow Chemical, 8, 37, 173–174, 175,
182–183, 184
Drinker, Cecil, 7
DuBois, Kenneth: and AMA Committee
on Pesticides, 102–103; classification of
organic phosphates, 105–106; on food
residues and contamination, 103–104;
on organophosphate toxicity, 189–191,
209–210; on potentiation, 180–181;
research on organophosphates, 92,
93–96, 108–110; and *Textbook of
Toxicology*, 154–155; and University of
Chicago toxicology program, 153–154
Dunbar, Paul, 9
Dunlap, Thomas, 188
DuPont's Haskell Laboratory, 8–9, 37
Dutch elm disease, 188

EDF v. Ruckelshaus, 189
Edsall, David, 7
Elixir Sulfanilamide, 18–26; AMA
concerns about, 19–20; analysis, 22–23;
deaths from, 3, 20–21, 22; development
of, 18–19; diethylene glycol as toxic
agent in, 19, 23–24, 30–32; FDA and
AMA analysis of, 21–23, 24–29, 31–32;
Geiling's method of analysis, 33–34; and
regulation of new drugs, 32–33, 34–37
Elliott, Michael, 195–200

256

INDEX

Elvolve, Elias, 15–16
endocrine disruption, 209, 217, 218
endrin, 159
England, 11, 98. *See also* Great Britain
environmental contamination, 58–65.
    See also wildlife
Environmental Defense Fund (EDF),
    188–189, 194
Environmental Health Perspectives, 212
environmental movement, ix, x–xi
Environmental Protection Agency
    (EPA), ix, 189, 192–194, 211, 213, 222
environmental science, xi
environmental toxicology, xiii
EPA. *See* Environmental Protection
    Agency (EPA)
EPN, 113–114, 202, 205, 207
ethylene glycol, 30
ethyl parathion. *See* parathion

Fairhall, Lawrence, 7
FDA. *See* Food and Drug Administration
    (FDA)
Federal Environmental Pesticide Control
    Act (FEPCA), 191–194
Federal Food, Drugs, and Cosmetics
    Act (FFDCA, 1938): Delaney Clause,
    150–151; limitations of, 116–117, 119;
    Miller Amendment, 149; passage of,
    36–37
Federal Insecticide, Fungicide, and
    Rodenticide Act (FIFRA, 1947):
    Congressional hearings and debate,
    116–120; deficiencies of, 167–168,
    191–194; provisions of, 120–121
Feichtmeir, Edmund F., 177–178
FEPCA (Federal Environmental
    Pesticide Control Act), 191–194
Fermi, Enrico, 84
FFDCA. *See* Federal Food, Drugs, and
    Cosmetics Act (FFDCA, 1938)
FIFRA. *See* Federal Insecticide, Fungicide,
    and Rodenticide Act (FIFRA, 1947)
fish and fish food, 38, 57–58, 60, 61–64, 198
Fish and Wildlife Service (FWS), 55–60
Fishbein, Morris, 20, 35, 36
Fisher, R. A., 26, 27

Fitzhugh, O. Garth, 26, 54, 105, 156
flame retardants, 212
Flannagan, John W., 119–120
Flemming, Arthur S., 150
fly bioassay, 53–54, 112–113
fonofos, 202, 203
Food and Drug Administration (FDA):
    and cancellation of aldrin and dieldrin,
    194; Cook's toxicity testing method-
    ologies, 110–113; and cranberry scare,
    150–151; DDT studies, 50–54; and
    Delaney Hearings, 145–146; Division of
    Pharmacology, 25–26, 29; and Elixir
    Sulfanilamide tragedy, 21–23, 26–29,
    31–32; and Federal Food, Drugs, and
    Cosmetics Act, 36–37; and ginger jake
    paralysis, 16–17; human tolerance
    levels, 164–165; insecticide toxicity
    studies, 99–101; and limitations of 1910
    insecticide law, 9–10, 12–13; and Miller
    Amendment, 149; and no-effect levels,
    174, 177; requests for industry testing of
    organophosphate potentiation, 172;
    and thalidomide, 175–176
Food Quality Protection Act (1996,
    FQPA), 211
food residues and contamination, 101,
    103–104; arsenates and, 10–13; Beech-
    Nut's near zero tolerance level in baby
    foods, 146–148; chlorinated hydrocar-
    bons, 100, 103–104, 133–134; cranberry
    scare, 150–151; Delaney Clause (1958),
    150–151; EPA cumulative risk assess-
    ment (2006), 211, 213; Hayes and Neal's
    statements on DDT safety (PHS),
    126–129; parathion, 102–103, 144–146,
    159–160; permethrin and, 197
Ford, William T., 31
Franck, James, 84
Frawley, John P., 113–114, 174–175,
    176–177, 181, 182
Freeman, Orville L., 166–168
Fuyat, Henry N., 156
FWS (Fish and Wildlife Service), 38, 55–60

gamma isomer, 101
Geigy, 39–40

257

INDEX

Geiling, E. M. K.: and analysis of new
drugs, 33–34; and antimalarial drugs,
76–83; and formation of University of
Chicago Toxicity Laboratory, 73,
74–75, 76; and radioactive digitoxin,
85–86; and Society of Toxicology,
156–157; and *Textbook of Toxicology*,
154–155; and toxicity studies of Elixir
Sulfanilamide, 23–24, 30, 32
general use pesticides, 192
Germany: development of sulfanilamide
in, 18; occupational medicine in, 6;
organophosphate research, 92–93
ginger jake. *See* Jamaica ginger
Goldfain, Ephraim, 15
Granger, Walter K., 118–119
Great Britain, 6, 66–68. *See also* England
Griffin, E. L., 119
Grob, David, 96–98, 102–103, 105
Gross, Harry, 14–15, 17
growth rates, 52
gypsy moth control, 4

Hamilton, Alice, 6–7
Harness, T. R., 84
Hartzell, Albert, 102–103
Harvard Medical School lead study, 7
Haskell Laboratory, 8–9
Hayes, Wayland J., Jr., 68–69, 126–129
Hazleton, Lloyd W., 106–108
Hazleton Laboratories, 106–108
Heal, Ralph, 146
Hedrick, E. M., 125, 133–134
Herber, Lewis, 157
Hercules Powder Company, 174
hexaethyl tetra phosphate (HETP), 92,
94, 99, 100, 104
Higgins, Elmer, 58–59
Hilts, Philip J., 36
Hitchner, L. S., 117–119
Holland, Emily G., 106–108
honeybees, 43, 46, 198
Horan, Walter, 131–132
houseflies, 45, 53–54, 131–132, 196–197
Howard, Leland Ossian, 40
Hub Products Corporation, 14–15
Hueper, Wilhelm C., 8–9, 137–139

Humphrey, Hubert, 166
Hutchinson, K. T., 122
hydrolization, 97, 102, 103, 104, 111, 114
Hygeia, 35

I. G. Farbenindustrie, 18, 77
imidacloprid (IMI), 221, 222
ingestion, 47–48, 51–52. *See also* food
residues
inhalation experiments, 65–66
Insecticide Act (1910), 2, 5–6, 25
insecticide industry. *See* chemical industry
insecticides/pesticides: agriculture's
increasing reliance on, xiii, 1–2, 3–4,
122, 132, 141–142, 163–164, 166–167;
amounts and types used, 13, 200–206,
206–209; classes of, xii–xiii, 171–172,
221–222 (*see also* carbamates; chlorinated hydrocarbons; organophosphates; systemic insecticides); education
of public about, vi, xi, 157–188, 168, 170;
hearings (*see* Delaney Hearings; Federal
Insecticide, Fungicide, and Rodenticide Act (FIFRA, 1947); Subcommittee
on Reorganization and International
Organizations hearings on interagency
coordination in environmental hazards);
industry development and testing
procedures, 182–184; labeling, 1, 5,
32–33, 117, 120–121, 214; Lehman's
(FDA) comparative studies, 99–101;
registration, ix, 117–121, 167–168, 191–194;
restricted use and general use categories, 192; risk assessment, xiii–xiv, 122,
163–165, 211, 217–220, 222–223. *See also*
legislation; *specific insecticides*
Institute of Toxicology, 179–180
Interagency Coordination in Environmental Hazards (Pesticides) Hearings
(Subcommittee on Reorganization and
International Organizations) (1963):
165–179; Carson's recommendations,
168–171; Freeman (Secretary of
Agriculture) on insecticide benefits *vs.*
risk; industry testimony on new-
pesticide development and testing,
182–184; Jaworski (Monsanto) on

258

INDEX

toxicity of natural products, 178–179; Johnson (Dow Chemical) on organophosphates, 173–174; Larrick (FDA) on organophosphate potentiation, 172; Lehman on toxicology-pharmacology gap, 179–180; Randolph (University of Michigan Medical School) on sensitization to organophosphates, 172; testimony about no-effect levels, 174–178, 179; testimony on need for interagency coordination, 169; testimony on potentiation, 180–181; Vos (FDA) on determining tolerance levels, 174; West (California Department of Public Health) on acute toxicity of organophosphates, 172–173

See Subcommittee on Reorganization and International Organizations hearings on interagency coordination in environmental hazards (1963)

International Conference on Alternative Insecticides for Vector Control (1970), 195–196

Jacobsen, Leon, 83
Jacobson, Martin, 166
Jamaica ginger and jake leg paralysis (ginger jake, jake leg), 2, 13–18, 161
Jaworski, Ernest J., 178–179, 183–184
Jeffries, Zay, 84
Jeffries Committee, 84
Johns Hopkins University School of Medicine, 96–98
Johnson, Julius E., 173–174, 175, 176–177, 182–183
joint toxicity/potentiation, 80–82, 83, 113–114, 160, 172, 180–181
Journal of the American Medical Association, 19–20, 69, 128
Jungle, The (Sinclair), 1

Kallet, Arthur, 10, 12, 13, 17
Kay, Gwen, 37
Keefe, Frank B., 119
Kelsey, Frances Oldham, 73, 80, 175–176
Kennedy, John F., 163
Kinkela, David, 41

Kleinfeld, Vincent A., 121; questioning of Bishopp (USDA), 142–145; questioning of Biskind, 126; questioning of Bromfield, 135; questioning of Decker, 132–133; questioning of Hayes and Neal (PHS), 127–129; questioning of Hueper, 137–138; questioning of Lehman (FDA) on parathion, 145–146
Klumpp, Theodore G., 31
Knipling, Edward F., 40–41, 41–42, 139, 143–144, 166, 167

labeling, 1, 5, 32–33, 117, 120–121, 214
lady beetles, 43
Lamb, Ruth deForest, 10
Langston, Nancy, 150, 175, 184–185, 209
Larrick, George, 172, 176
Laug, Edwin P., 25, 26, 27–28, 29, 53–54, 112, 113
LD_{50} (Lethal Dose 50): and additive effects (potentiation), 113–114; for carbamates, 109; comparative studies of, 106, 197; database, 89; for DDT, 47, 53, 134; development of, 27–29; for neonicitinoids, 221; and no-effect level, 177; for organophosphates, 95, 96, 99, 106, 113–114, 173, 196, 219; for pyrethroids, 196, 197
lead arsenate, 4–5, 9, 11–12, 13, 54
lead industry and lead poisoning, 6–7
legislation: Elixir Sulfanilamide tragedy and regulation of new drugs, 32–33, 34–37; Federal Environmental Pesticide Control Act (FEPCA), 191–194; Federal Food, Drugs, and Cosmetics Act (FFDCA, 1938), 36–37, 116–117, 119, 149; Federal Insecticide, Fungicide, and Rodenticide Act (FIFRA, 1947), 116–120, 120–121, 167–168, 191–194; Food Quality Protection Act (1996, FQPA), 211; Insecticide Act (1910), 2, 5–6, 25; National Environmental Protection Act (1970), 189; overview, 214–217; Pure Food and Drug Act (PFDA, 1906), 1–2, 5, 17–18, 31, 33, 35–36

259

INDEX

Lehman, Arnold J.: and AMA Committee on Pesticides, 102–103; as chief of FDA Division of Pharmacology, 105; insecticide toxicity studies, 99–101; on parathion, 145–146; and Society of Toxicology, 156; and *Toxicology and Applied Pharmacology,* 155
Lehmann, Karl, 6
Lethal Dose 50, see LD_{50}
Libby, W. F., 87
Lightbody, Howard, 25
Lillie, R. D., 48, 51
lindane, 147, 168, 173, 204
liver pathology, 48, 51
Long, Perrin H., 32
Lushbaugh, C. C., 83–84
Lyndol, 14–15

MacDougall, Dan, 105
malaria, 44–45, 76–83, 141
malathion (malathon): amounts used, 200, 201, 203, 204, 205, 207, 208; Carson on, 160; occupational disease attributed to, 173; toxicity, 106, 108–109, 113–114, 197
mammals: Carson on, 158–160; and chlorpyrifos, 203; and DDT, 55, 59, 61, 64–65; and HETP, 94; and OMPA, 95; and organophosphates, 105–109, 112, 173; and parathion, 95; and pyrethroids, 195, 196–197, 199–200
Mancuso, Thomas, 138
Mangun, George, 94–95
Markowitz, Gerald, 184–185
Massengill Company, 19–26, 30–31
McCaulley, D. F., 112–113
McCloskey, W. T., 25
McLean, Franklin D., 74–75
Mellon Institute, 109
methyl parathion, 112, 200, 201, 202–203, 204, 207, 208
Mexican fruit fly, 44
milk contamination, 54–55, 100, 103–104, 112, 125, 129, 142–144
Miller, A. I., 121, 124, 130–131, 134, 139, 143, 145, 149
Miller, Lloyd C., 25
Miller Amendment (1954), 149, 151

monocrotophos, 210, 211
Monsanto Chemical Company, 178–179, 183–184
Morris, Harold P., 25, 26, 136–137
Morris, Herman, 25
Moulton, F. C., 4
Müller, Paul, 39–40
Mulliken, R. S., 84
Murphy, Sheldon, 109–110
muscarine, 97, 159
Myers, C. N., 11–12

naphthalene, 18
Nash, Linda, 211
National Academy of Science (NAS), 194
National Association of Insecticide and Disinfectant Manufacturers, 34, 35
National Cancer Institute (NIH), 136–138
National Defense Research Committee (NDRC), 73–74
National Environmental Protection Act (1970), 189
National Farmers Union, 119
National Institutes of Health (NIH), 47–50, 65–66, 79, 136. *See also* U.S. Public Health Service
National Pest Control Association, 146
Neal, Paul A., 48–50, 65–66, 120, 126–128
Nelson, Arthur, 105
Nelson, Dallas, 105
Nelson, Erwin, 25–26
neonicitinoids, 221–222
neurotoxicity/central nervous system effects: DDT and, 48, 53, 65, 67; organophosphates and, 92–98, 106, 107, 110–113, 159, 191, 209–210, 211, 212. *See also* cholinesterase inhibition
Newell, S. R., 116–117
New Jersey health officials, 123
New Yorker, 162, 163
New York Times, x, 22, 34–35
nicotine, 198–199, 221
nicotinoids, 221
NIH. *See* National Institutes of Health

260

INDEX

nitrogen mustards, 83–84
no-effect levels, 164–165, 174–178, 179

occupational exposure and illness,
   6–9, 37, 97, 98, 160–161, 172–173,
   201, 211
Oettingen, Wolfgang F. von, 8, 30–31,
   48–50, 65–66, 156
oil sprays (paraffinic oils), 118, 200, 204,
   207, 208
Oklahoma Experimental Station, 143
OMPA (Pestox III, Octamethyl Pyrophosphoramide), 95–96, 104, 106
*100,000,000 Guinea Pigs: Dangers in
   Everyday Foods, Drugs, and Cosmetics*
   (Kallet and Schlink), 10, 12, 17, 182
Oreskes, Naomi, 184–185
organochlorines. *See* chlorinated
   hydrocarbons
organophosphates (organic phosphates):
   AMA Committee on Pesticides'
   recommendations, 102–103; amounts
   and types used, xi, xii, xiv, 200–206; as
   biocides, 219–220; and cancer, 209;
   Carson on, 159–161; children and, 211,
   212; and chlorinated hydrocarbons
   trade-off, 140–141, 185–186, 187; and
   cholinesterase inhibition (*see* cholinesterase inhibition); classification of,
   105–106; compared to DDT, 99–101;
   Cook's toxicity testing methodologies
   (FDA), 110–113; DuBois on toxicity of,
   189–191; EPA cumulative risk assessment
   of (2006), 211, 213; high acute toxicity
   of, 98, 102, 171–172, 173–174, 187, 209,
   219–220; hydrolization of, 102–103;
   Lehman's (FDA) toxicity studies of,
   99–101; neurotoxicity and chronic
   exposures, 92–98, 106, 107, 110–113, 159,
   191, 209–210, 211, 212; and potentiation,
   113–114, 172; as replacements for DDT,
   188, 189–191, 216–217, 218–220; University
   of Chicago Toxicity Laboratory research
   (DuBois and colleagues), 92–96,
   103–104, 105–106, 108–110; and wildlife,
   161, 210–211. *See also specific organophosphates*

Our Children's Toxic Legacy (Wargo), 211
oysters, 168

paired samples *vs.* random samples, 62
Palm, Charles E., 131–132
Paracelsus, xiii
paraffinic oils (oil sprays), 118, 200, 204,
   207, 208
paraoxon, 174
parathion (ethyl parathion): Carson on,
   159–161; and cholinesterase inhibition,
   95, 112–114; cumulative action of, 100;
   Delaney Hearings on use and toxicity
   of, 125–126, 144–146; DuBois and
   Coon on toxicity of, 105–106;
   environmental persistence, 165, 171;
   and food residues, 97–98, 102–103, 144–
   146, 159–160; Hazleton and Holland on
   toxicity of, 106–109; and houseflies,
   196–197; Johns Hopkins University
   research on symptoms of exposure,
   96–98; Lehman on toxic exposure
   levels,100, 99, 100, 101, 102–103; and
   milk contamination, 100, 112; occupational illnesses and deaths, 96–97, 98,
   102, 173–174; as substitute for DDT,
   190, 191; use of, 1966–1989, 200, 201,
   202, 203, 204; use of, 1989–2002, 207,
   208; and wildlife, 208
Paris green, 4, 5, 11
Patuxent Research Refuge wildlife
   studies, 59–61
Perkins, John, 39–40, 149
permethrin, 197
persistence, 196–197, 222
Pesticide Action Network, x
Pesticide Analytical Manual, 113
pesticide industry. *See* chemical
   industry
PFDA. *See* Pure Food and Drug Act
   (PFDA, 1906)
pharmaceutical regulation, 175–176
pharmacology, xiii, 25, 153, 154, 156,
   179–180
phorate, 201, 202, 207, 208
phosphoramides, 105, 106
Plzak, Vivian, 89

261

INDEX

potentiation/joint toxicity, 80–82, 83,
  113–114, 160, 172, 180–181
precautionary principle, 55, 131, 147, 178
pregnancy, drug therapies and, 87
prenatal exposures, 212
President's Science Advisory Committee
  (PSAC) report, 163–165, 171–172
primaquine, 79
Princi, Frank, 130–131
Proctor, Robert, 9, 138
protest registrations, 167–168
PSAC. *See* President's Science Advisory
  Committee
public education, vi, xi, 157–188, 168, 170,
  187–188
Public Health Service. *See* U. S. Public
  Health Service
Pure Food and Drug Act (PFDA, 1906):
  inadequacies of, 17–18, 31, 33, 35–36;
  passage, 1–2, 5; provisions, 5
pyrethrins, 101
pyrethroids, synthetic, 195–200. *See also*
  specific synthetic pyrethroids
pyrethrum, 4, 195

radioactive digitoxin, 85–89
radioactive nicotine, 85
radioisotopes, 84–89
Randolph, Theron G., 172
random samples *vs.* paired samples, 62
registration for a pesticide, ix, 117–121,
  167–168, 191–194
resistance, 42–43, 44, 131–133, 198, 202
restricted use pesticides, 192
Ribicoff, Abraham, 166, 168–171, 174,
  176–178, 184
Ribicoff subcommittee. *See* Subcommittee on Reorganization and International Organizations hearings on
  interagency coordination in environmental hazards (1963)
Richardson-Merrell Company, 175–176
risk assessment, xiii–xiv, 122, 163–165,
  211, 217–220, 222–223. *See also specific*
  insecticides
Rockefeller Foundation, 41
Root, Mildred, 89

Rosner, David, 184–185
Ross, Benjamin, 138
rotenone, 101, 199
Rothamsted Experimental Station
  (England), 195–197
Ruckelshaus, William, ix, 189
Rudd, Robert, 157
Russell, Edmund, 69

Sanderson, E. Dwight, 5
Savannah River National Wildlife
  Refuge, 61–65
Schlink, F. J., 10, 12, 17
Schrader, Gerhard, 92, 95–96
Schueler, F. W., 82–83
Science, 142
scientific uncertainty, 121, 129–133,
  148–149, 185, 222
Sellers, Christopher, 6, 7–8
S. E. Massengill Company, 19–26, 30–31
sensitization, 172, 181
Sevareid, Eric, 162–163
Sevin. *See* carbaryl
Shaw, W. C., 139
Shell Chemical Company, 194
Shell Development Company, 177–178
Silent Spring (Carson): and American
  environmental movement, ix, x–xi,
  212; and DDT (chlorinated hydrocar-
  bons), 157–159, 161, 187–188; indictment of industry and government,
  xi–xii, 153, 182, 200; on limited use
  of pesticides, 161–162, 169, 200; on
  organophosphates, 159–161, 187,
  219–220; overview, x, 157–162;
  response to, xii, 182, 186–188,
  222–223
Sinclair, Upton, 1
Smith, Maurice, 15–16, 47–48
Smith, Russell, 119
Society of Toxicology, 156–157
soil contamination, 46, 125, 161, 165, 171,
  200, 201
specificity concept, 58, 100
Spendlove, George A., 123
Spurr, Charles, 83
state sovereignty, 118

INDEX

Statesville Hospital malaria program, 78–80
Stickel, Lucille, 57
Stohlman, Edward F., 47–48
Stone, R. S., 84
Subcommittee on Reorganization and International Organizations. *See* Interagency Coordination in Environmental Hazards (Pesticides) Hearings
sulfanilamide, 18–19. *See also* Elixir Sulfanilamide
Surber, Eugene, 57–58
Swainson's hawks, 210
synergism, 80–82, 89, 160, 165, 172
systemic insecticides, 95–96, 104, 201, 221–222

Tarzwell, Clarence M, 61–65
Tennessee Valley Authority (TVA), 44–45
TEPP (Tetraethyl Pyrophosphate), 97, 99, 101, 102, 104, 105
terbufos, 202, 203, 204, 205, 206, 207, 208
termites, 42
Tetraethyl Pyrophosphate (TEPP), 97, 99, 101, 102, 104, 105
Texas Research Foundation, 133–135
Textbook of Toxicology (Geiling and DuBois), 154–155
thalidomide, 174–176
therapeutic index, 82–83
thiametoxam, 222
thionophosphates, 108–109
Thomas, C. A., 84
Throne, Binford, 11–12
Time, 38
tobacco, 42–43
TOCP (triorthocresyl phosphate), 14–18, 161
tomato-fruit worm outbreak (1945), 141
toxaphene, 200, 201, 202
toxicity: Cook's testing methodologies, 110–113; dosage-mortality curve, 26–28; genetic effects vs, 169–170; LD$_{50}$, 27–29; specificity concept, 58; "toxicity of natural products" defense, 178–179. *See also* acute vs. chronic toxicity; *specific insecticides*

toxicology, history of: consolidation of, as discipline, 153–157; creation of Institute of Toxicology, 179–180; DDT studies and, 70–71; Elixir Sulfanilamide investigation and, 23–29; first corporate laboratories, 8–9; genetic toxicology, 170; Hueper and, 8–9; and industrial hygiene, 6–7; industrialization and, 1–6; Society of Toxicology, 156–157; *Toxicology and Applied Pharmacology,* journal, 155–156; University of Chicago Toxicity Laboratory and, 72–73, 89–90. *See also* environmental toxicology
Toxicology and Applied Pharmacology, 155–156
Tox Lab. *See* University of Chicago Toxicity Laboratory (Tox Lab)
Train, Russell, 194
Treichler, Ray, 56–57
triorthocresyl phosphate (TOCP), 14–18, 161
trypanosomes, 82–83
Tugwell, Rexford, 9, 12–13
typhus, 41–42

uncertainty. *See* scientific uncertainty
Union Carbide, 109
United States Pharmacopoeia (USP), 13–14
University of Chicago Department of Pharmacology, 23–24, 73, 76
University of Chicago Toxicity Laboratory (Tox Lab): analysis of nitrogen mustards, 83–84; antimalarial drug therapies research program, 76–83; chemical warfare agent evaluations, 73–76; formation and development of, 72–73; organophosphate research, 92–96, 105–106, 108–110; radiation research, 153–154; radioisotope research, 84–89
U.S. Air Force Radiation Laboratory, 153–154
U.S. Army, 72, 96

263

INDEX

U.S. Army Sanitary Corps, 41
U.S. Congress. *See* Delaney Hearings;
  Interagency Coordination in Environmental Hazards (Pesticides) Hearings
  (Subcommittee on Reorganization
  and International Organizations);
  legislation
USDA: and administration of FIFRA,
  117–121, 119–120; Bishopp (Bureau of
  Entomology and Plant Quarantine)
  on risks of DDT, 139–145; chemical
  industry capture of, 146, 185; chlordane
  recommendations, 146; and cranberry
  scare, 150–151; criticism of Agricultural
  Research Service, 185; and DDT
  analysis, 40; and Delaney Hearings,
  122; and FIFRA hearings, 116–117;
  and Insecticide Act of 1910 conflict of
  interest, 5–6, 25; recommendations
  on DDT use on/near dairy animals,
  142–144; relationship to chemical
  companies, 146
"Use of Pesticides, The." *See* President's
  Science Advisory Committee (PSAC)
  report
U. S. Fish and Wildlife Service (FWS),
  55–60
U.S. Public Health Service (PHS): and
  DDT toxicity to humans, 68–69; and
  examination of insecticides, 29; and
  ginger jake paralysis, 15–16; Hayes and
  Neal's testimony at Delaney Hearings,
  126–129; Savannah River National
  Wildlife Refuge DDT larvaciding
  studies, 61–65
Utah Department of Health, 123

velvetbean caterpillar outbreak (1946), 141
Vos, Bert, 105, 174

Wallace, Henry A., 32–33
Ward, J. C., 102–103
Wargo, John, xii, 120–121, 149, 150,
  191–192, 193, 211
Watkins, Harold Cole, 19, 30–31, 32
Weigele, Carl E., 123
West, Irma, 172–173
Whitaker, Adelynne, 2, 5, 120
White-Stevens, Robert, 163
Whorton, James, 3–4, 11
Wiesner, Jerome, 163
wildlife: Carson on insecticides and,
  159–162; and DDT ban, 212; Fish and
  Wildlife Service studies, 55–58; growing
  concerns about DDT and, 187–188;
  neonicitinoids and, 222; and organo-
  phosphates, 210–211; and parathion,
  161; Patuxent Research Refuge studies,
  59–61; President's Science Advisory
  Committee recommendations regarding
  study of, 165; Savannah River National
  Wildlife Refuge studies, 61–65;
  Tennessee Valley Authority (TVA)
  studies, 44–45; terbufos and chlorpyrifos and, 203; wild animal laboratory
  experiments, 55–58. *See also* birds
Wiley, Harvey, 5
Williams, Martin W., 156
Wolfe, Humphrey D., 8
Woodard, Geoffrey, 26, 29
workers. *See* occupational exposure and
  illness
World War II: chemical warfare research,
  73–76; and malaria, 76–77
World Wildlife Fund, 209

Yamamoto, Izuru, 221

Zeidler, Othmar, 39